Workbook for Geologic Catastrophes

Second Edition

Carla Whittington

Eric Baer

Highline Community College

KENDALL/HUNT PUBLISHING COMPANY
4050 Westmark Drive   Dubuque, Iowa 52002

Cover graphs courtesy of USGS.

Copyright © 2005, 2006 by Carla Whittington and Eric Baer

ISBN 0-7575-2632-2

Kendall/Hunt Publishing Company has the exclusive rights to reproduce this work,
to prepare derivative works from this work, to publicly distribute this work,
to publicly perform this work and to publicly display this work.

All rights reserved. No part of this publication may be reproduced,
stored in a retrieval system, or transmitted, in any form or by any
means, electronic, mechanical, photocopying, recording, or otherwise,
without the prior written permission of the copyright owner.

Printed in the United States of America
10 9 8 7 6 5 4 3 2 1

Preface and Acknowledgements

We are indebted to all the people that have taken this work from concept to the actual book. Thank you to all the students who have tried and made suggestions on all of these exercises as they were developed. Many faculty made suggestions, reviewed pieces or even the entire manuscript, even suggested the basic outlines of the assignments we have included, or authored works that inspired this work. These include (in no particular order) Lisa Gilbert (Williams-Mystic Maritime Studies), Eriks Puris (Portland Community College), Emanuela Baer (Shoreline Community College), Duncan Foley (Pacific Lutheran University), Gina Erikson (Highline Community College), Larry Braille (Purdue University), Hobart King (Mansfield University), James Loettle (Highline Community College) and the Geologic Society of America. We thank them for their inspiration and assistance. All errors, are, of course our own.

The exercises use data, websites, and information from a variety of sources including the United States Geologic Survey, the Pacific Northwest Seismic Network, the Federal Emergency Management Agency, The Burke Museum, King County government, and other governmental agencies. We thank those who work for these agencies to increase our knowledge and understanding of natural hazards and for making their information available to all.
Thank you also to Kristina Stolte and other staff at Kendall-Hunt for their assistance in developing these and other student oriented publications.

About the Authors

Carla M. Whittington earned a Bachelors of Science from Indiana-Purdue University at Fort Wayne in 1993 and a Masters of Science from Indiana University in 1996. Her specialty area is igneous petrology. Carla enjoys traveling, backpacking, and the volcanoes of the Cascades. She has taught at Highline Community College since 1998.

Eric M. Baer has a Bachelors degree from Carleton College in Minnesota and a Ph.D. from the University of California, at Santa Barbara. He has studied volcanic rocks in Ecuador and Southern Japan. Eric enjoys backpacking, traveling, and eating. He has taught at Highline Community College since 1997.

Financial Disclosure

The authors will not receive royalties for its sale or adoption at Highline Community College

Table of Contents

1. Introduction

1.1	Introduction to Class Materials	1
1.2	Sediments and Rocks!	3
1.3	FEMA and the Government's Role in Disasters	17
1.4	Introduction to Library Resources	19
1.5	Video: Plate Dynamics	21
1.6	Maps – Where Do You Live and What Do You Live On?	25
1.7	Plate Tectonic Setting of the Pacific Northwest	29
1.8	Support for You!	31
1.9	What is Plagiarism?	33

2. Earthquakes

2.1	Identifying Faults	35
2.2	Earthquake Information on the Internet	37
2.3	Epicenters and Magnitude	39
2.4	Calculating Earthquake Magnitude	49
2.5	Seismic Design	51
2.6	Shaking during the 2001 Nisqually Earthquake	53
2.7	Earthquakes in the Pacific Northwest	71
2.8	Ground Shaking and Damage at Your House	73
2.9	Earthquake Hazards at Your House	77
2.10	Forecasting Future Earthquakes	81
2.11	The Seattle Fault Scenario	91
2.12	Finding Patterns in Earthquakes	95
2.13	Video : Cascadia –The Hidden Fire	99

3. Volcanoes

3.1	Volcanic Hazard Maps	101
3.2	Volcano Movie Review	105
3.3	Disaster in Armero	107

3.4	Ash Fall Hazards at Your Home	109
3.5	Video: Understanding Volcanic Hazards	113
3.6	Video: Kilauea, Close Up of an Active Volcano	115
3.7	Video: Perilous Beauty	117
3.8	Video: Kilauea, Lava Flows	121
3.9	Phases of Eruption of Vesuvius, 79 AD	123
3.10	Monitoring of Volcanoes	127

4. Landslides

4.1	Angle of Repose	129
4.2	Forces in Mass Movement	135
4.3	Examining Slope Stability	139
4.4	The Liquid Earth	145
4.5	Landslides in the Puget Sound	147
4.6	Mass Movement Hazards Around Your Home	151
4.7	Video: Mass Wasting	155

5. Flooding

5.1	Flood Frequency in Seattle Area	157
5.2	Flooding in Your Neighborhood	165
5.3	Video: Running Water	167
5.4	Video: Running Water - Landscape Evolution	169

6. Reference Materials

6.1	Reference Style Guide for Paper/Assignments	173

1.1 Introduction to Class Materials
(Textbook)   10 points

Name _____

Due Date _____

The textbook is full of useful information that supplements this course. In addition to general information about the hazards that we cover in class, the textbook also has specific information about hazards in the Pacific Northwest. Examples of this include earthquakes (p. 88-89) tsunami (118-122) and volcanoes (p. 128-134 and 157-161).

Like many science textbooks, geology textbooks use graphs, maps, cross-sections and pictures to convey important information, data, and concepts. Students, who often focus on reading the text, have a tendency to overlook these graphics.

The correctly formatted bibliography for your textbook is:

> Hyndman, D. and Hyndman, D., 2006, *Natural Hazards and Disasters*; Thomson Brooks/Cole, California, pp. 420.

Find Figure 2.13 on page 20 of the text.
This graphic shows a schematic "cross-section" from the surface of the Earth, down to a depth of 225 km. Notice that the Earth is composed of many physical and chemical layers.

1. The crust of the Earth is the very thin layer of low-density rock at the surface and is divided into two types: oceanic and continental. Which type of crust is the thickest? _____

2. The lithosphere is a physical layer of the Earth that is very strong and solid. It is made of the crust and the upper part of the mantle. How deep (approximately) does the lithosphere extend down to? _____

Find Figure 5-30 on page 117 of the text.
This graphic shows tsunami travel times across the Pacific Ocean.

3. If an earthquake occurred off the coast of Chile, how long would it take a tsunami wave to hit Hawaii? _____

1.1 Introduction to Class Materials
(Textbook)   10 points

Find Figure 2.18 on page 24 of the text.
The map conveys information about the plate tectonic setting of the Pacific Northwest. In Washington, we live along a boundary between two tectonic plates.   As the plates collide, one plate is "subducted" beneath the other, meaning the plate is forced down into the Earth's mantle where it is ultimately destroyed.

4.   What are the names of the two plates
       that are involved in this collision off
       the coast of Washington?

5.   What is the name of our local
       subduction zone?

Find Figure 7.1 on page 152 of the text.
This diagram compares the different types of volcanoes in terms of their size and volume.

6.   What type of volcano is Mauna Loa?

7.   Mt. Rainier is a typical stratovolcano.
       Which is bigger, Mt. Rainier or Mauna
       Loa?

Find Figure 7.16 on page 158 of the text.
This diagram shows the major Cascade Range volcanoes and their eruptions through the last 4,000 years.

8.   How many times has Mt. Rainier
       erupted within the last 2000 years?

9.   Has Mount Rainier erupted within the
       last 200 years?

Find Figure 7-41 on page 173 of the text.
This graph shows the relative magnitude of selected volcanic eruptions.

10.   How does the volume of erupted
       material compare between Mt. St.
       Helens 1980 eruption and the
       eruptions of Yellowstone Caldera?

1.2 Sediments and Rocks!
(Rock/Sediment Samples in Class)   15 points

Name _____

Due Date _____

This activity gives you an opportunity to look at sediments and rocks. Let's start with some definitions:

Sediment: Solid fragmental material that is transported and deposited by air, water, or ice, or that accumulates by other natural agents, such as chemical precipitation from solution or secretion by organisms, and that forms in layers on the Earth's surface at ordinary temperatures in a loose, unconsolidated form; e.g. sand, gravel, silt, mud, clay, alluvium, till, loess.

Rock: A consolidated aggregate of one or more minerals, (e.g. granite, shale, marble) or a body of solid organic material (e.g. coal).

> Bates, R.L. and Jackson, J.A., 1987, **Glossary of Geology, 3rd Edition**, American Geological Institute, Alexandria, pp. 788.

Looking at Sediment

Most sediment is formed because rocks exposed at the surface of the earth decay and break down (weather) over time. Some sediments have a biologic origin. Others form when minerals form in saline waters and sink to lake bottoms or the ocean floor.

At the surface of the Earth, blowing wind, moving water, flowing ice and even gravity (!) can erode these sediments and transport them to different locations where they get deposited and accumulate in loose piles. Loose (unconsolidated) sediments are named based on the predominant particle size in the deposit. Here is a table showing how sediments are classified and named:

Figure 1. Particle Size Classification of Sediments		
Particle Size (mm)	**Name of Loose Particle**	**Name of Loose Sediment**
Coarse > 256	Boulder	
64 - 256	Cobble	Gravel
4 - 64	Pebble	
2 - 4	Granule	
0.063 - 2	Sand	Sand
0.004 - 0.063	Silt	Mud
Fine < 0.004	Clay	

1.2 Sediments and Rocks!
(Rock/Sediment Samples in Class)   15 points

You will be visiting different stations, examining the samples there, and responding to questions. Stations 1-4 contain sediment samples. Stations 5-8 contain rock samples. Hand lens and rulers will be available for your use.

Please try not to spill the sediment samples.   Some are irreplaceable (Mount St. Helen's may not erupt again for hundreds of years)! Others are simply difficult to obtain (I don't get to Hawaii that often)!

1.2 Sediments and Rocks!
(Rock/Sediment Samples in Class)   15 points

Station 1.   Sediments: Let's Play in the Dirt!

Examine the three samples A, B, and C. and answer the following questions. You can use a ruler to measure the particles and compare them to the chart if you are unsure.

1. What is the predominant grain-size in each of the three samples?

_____

2. Which samples are well-sorted? (The particles in the sample are all roughly the same size.)

_____

3. Which samples are poorly sorted? What other particle sizes do you see in the sample?

_____

4. Are the particles in any of the samples rounded? If so, which?

_____

5. Are the particles in any of the samples angular? If so, which?

_____

6. When sediments are first formed from weathering of rock, the particles are often very angular. How might particles become rounded?

_____

7. Can you suggest some natural environment in which particles of this size would accumulate or be deposited? If so where?

_____

5

1.2 Sediments and Rocks!
(Rock/Sediment Samples in Class)   15 points

Station 2.   Sediments: Let's Play in the Sand!

Samples A – E are all sand. Remember, sand doesn't refer to any specific composition; it is a name for a particle of a certain size. Take a look at them to see how much sand can vary! Samples A – E are all sands that were collected in Hawaii. Sample E comes from a different location. Examine the samples and answer the following questions.

1.	Which sample is the coarsest (has the largest particle size)?
2.	Which sample is the finest (has the smallest particle size)?
3.	Which sample has the poorest sorting? (It has a combination of larger and smaller sand particles).
4.	Wind-blown deposits of sand are typically very well-sorted and very-well rounded and quite fine. Which of the sands here do you think are the result of wind transportation and deposition?
5.	Examine sample C closely. You can use a hand lens if you want. What do you think the sand particles are composed of? Where might you find sand of this composition accumulating?
6.	Which of the sands do you find the most unusual and why?

1.2 Sediments and Rocks!
(Rock/Sediment Samples in Class)   15 points

Station 3.   Sediments: Let's Play in the Ash!

Samples A and B are both ash deposits from the May 18th, 1980 eruption of Mount Saint Helen's in Washington State. Sample C is from a different location (I don't know where!)

Examine the samples and answer the following questions.

---

1. What is the predominant size of the particles at this station? Remember to consult your particle size classification!

Sample B from the eruption of Mount St. Helens is one the very fine end of the size range classification. It has the consistency of flour. It was collected in Eastern Washington south of Spokane.

Sample A, also from Mount St. Helens was collected near Yakima, just east of the Cascades.

---

2. Can you think of an explanation for why the ash collected 200 miles from the volcano would be finer in size than the ash collected much closer to the volcano? (Please give an explanation.)

1.2 Sediments and Rocks!

(Rock/Sediment Samples in Class)   15 points

Station 4.   Sediments: Let's Play in the Mud!

The sample at this station is clay! In fact, it is pottery clay; the substance used to make ceramic dishware and art.

Clay sized particles can be of many varied compositions. Most clays can be used in one industrial processes or another, depending on the properties of the clay. This makes them an important economic resource. Some clays are extremely absorbent. These clays are useful in cleaning up oil spills, removing ink from paper during recycling, and even absorbing moisture and odor in kitty litter! Other clays are used in disbursing the color in paints or lubricating the machinery during the drilling of wells, and as a binder in processed foods. (Yes, you eat "dirt" on a fairly regular basis...) With literally thousands of uses, clays just may be one of the most important economic resources we have!

Examine the sample and answer the following questions.

1.   Look again at the classification chart for particle size. Is there anything unusual about this sample? What is it?

2.   Try picking up one of the chunks in the sample. What happens when you crush it between your fingers? (You might want to try this with one of the smaller chunks!)

3.   Give a suggestion as to what may be causing the clay particles in the sample to stick together.

4.   Do you (can you?) see the clays are individual particles, like those of silt or sand, or gravel?

1.2 Sediments and Rocks!
(Rock/Sediment Samples in Class)   15 points

Looking at Rock

Rocks are strong and hard because the mineral crystals, particles, or grains that make up the rock are stuck together into a consolidated mass.

Rocks are classified into three main groups. These groups are delineated on the basis of how they were formed.

Igneous Rocks form from magmas and lavas that cool and solidify.

Sedimentary rocks form from sediments that are cemented together by natural processes.

Metamorphic Rocks form when igneous or sedimentary rocks recrystallize in response to increases in temperature and/or pressure.

At stations 5 – 8, you will have the opportunity to look at the different groups of rocks.

1.2 Sediments and Rocks!
(Rock/Sediment Samples in Class)   15 points

Station 5.   Crystalline and Clastic Rocks

There are two basic ways that the pieces get "stuck" together, so rocks can be divided into two main types: **crystalline** and **clastic**.

Crystalline rocks: When some rocks form, the mineral crystals that comprise the rock begin to grow. The growing minerals get in each other's way because the space is limited. When several types of minerals are growing at the same time, they tend to grow into and interlock with each other. The individual crystals usually have irregular boundaries with no space between them and the surrounding crystals. There rocks are called crystalline rock. Most igneous and metamorphic rocks are crystalline.

Clastic rocks: When a group of distinct, individual particles or grains are solidified into rock, we call them clastic. The particles are usually sediment of some type and range in composition. The grains or particles have distinct edges and boundaries. Natural (usually clear) cements have formed in between the particles to "glue" them together. Most sedimentary rocks are clastic.

There are four rocks at this station. The names are **Granite, Diorite, Arenite**, and **Coquina**. Examine the rocks and answer the following questions.

1. Which of the four rocks appear to be crystalline rocks? (Give the names.)

---

2. Which of the rocks appear to be clastic? (Give the names).

---

3. What is the predominate grain or particles size in the arenite? (Use the particle size classification chart.)

---

4. Examine the particles that make up the coquina. What do you think the particles are comprised of?

1.2 Sediments and Rocks!
(Rock/Sediment Samples in Class)   15 points

Station 6.   Igneous Rocks

Igneous rocks form when magmas and lavas cool and solidify. As the liquid rock cools, mineral crystals begin to nucleate in the liquid and grow. They continue to form until the liquid is used up and only the solid rock remains. As the minerals form, they grow into each other in a random, interlocking pattern. Therefore igneous rocks are (mostly) crystalline.

Crystal growth is a slow process and it takes a long time for large crystals to develop. Therefore the sizes of the crystals in the rock tell us a bit about how slowly or quickly it took the liquid to solidify. Magmas deep in the Earth's crust cool down very slowly. The rock that surrounds the patch of magma doesn't conduct heat away very quickly, so crystals have lots of time to nucleate and grow to large sizes. Igneous rocks with lots of large visible mineral crystals of many different colors are called **intrusive** igneous rocks because their magmas never reached the surface of the Earth.

When magmas make it to the surface, they erupt as lavas. Lavas that come into contact with the atmosphere and/or water cool extremely fast. Many crystals will nucleate in the lava, but none of them get a chance to grow very large! By the time the lava is totally solidified, most of the crystals are still very fine and often we can only see them with a microscope! Igneous rocks with microscopic crystals are very fine-grained and uniform in color. Sometimes they may be *glassy*; other times they may have lots of *pores* (holes) in them. They are called **extrusive** igneous rocks because they form from lavas extruded on the surface of the Earth.

There are six igneous rocks at this station. There names are: **gabbro**, **granite**, **dacite**, **rhyolite**, **pumice** and **obsidian**. Examine them and answer the following questions.

1. Which of the rocks are intrusive igneous rocks? (Give the names).

_____

2. Which of the rocks are extrusive igneous rocks? (Give the names).

_____

3. The pumice has many pores in it, which makes it very light. Speculate on how you think the pores got there?

1.2 Sediments and Rocks!
(Rock/Sediment Samples in Class)   15 points

Station 7.   Sedimentary Rocks

Sedimentary rocks form from sediments. Sediments can be of any composition, but most sediments are pieces of minerals and rock that have been weathered off of preexisting rock material. Some sediment is biologic in origin. The sediments in sedimentary rock were once underlined. If these unconsolidated sediments are buried and compacted, natural processes can cement them together into consolidated rock.

To identify sedimentary rocks, we look at the predominate size of the sediment that makes up the rock. If there are large amounts of pebbles and cobbles in the rock, then it is called **conglomerate** or **breccia**. If the sediment is mainly sand-sized, we call the rock **sandstone**. Silt-sized sediment is the main component of siltstone. And rocks made of clay-sized particles are called **mudstone** or **shale**.

Examine the four rocks labeled A – D and answer the following questions.

1. Which of the samples contains pebble or cobble sized particles? (Give the letters).

_____

2. What other particle sizes can you identify in samples A and B?

_____

3. Compare the large particles in A and B. In one sample the pebbles are rounded, like they've been tumbles in a stream. We call this rock conglomerate. Which of the two samples is a conglomerate?

_____

4. Again compare A and B. In one sample the pebble sized pieces are very angular. We call this rock breccia. Which of the two samples is a breccia?

_____

5. What is the predominate particle size in sample C? What is the name of this rock?

_____

6. What is the predominate particle size in sample D? What is the name of this rock?

_____

12

1.2 Sediments and Rocks!
(Rock/Sediment Samples in Class)   15 points

Station 8.   Metamorphic Rocks

Metamorphic rocks form when other rocks (igneous and sedimentary) are subjected to higher pressures and/or temperatures. The higher temperatures and pressures make the minerals in the rock unstable, and the chemical components of the rock recombine to form new minerals that are stable at the higher pressures and temperatures. We call this process *recrystallization* and it occurs in a solid state (meaning that the rock does not melt during recrystallization).

Pressure has an odd effect during recrystallization. It can cause the minerals in the rock to realign and grow in a preferred orientation. If this happens, the rock develops a "layered" appearance because the minerals or grains are all lined up in the same direction. We call this preferential arrangement of minerals in the rock **foliation**.

Also, during recrystallization, new minerals that were not in the rock prior to metamorphism can form and become quite large. Garnet and many other precious gemstones form only during metamorphism.

The four metamorphic rocks at this station are: **schist**, **quartzite**, **slate**, and **gneiss**. Examine the rocks and answer the following questions.

1.   Which of the samples exhibit foliation (layering)? Give their names.

_____

2.   What of the samples do not have foliation? Give their names.

_____

3.   Which sample contains garnet? Give its name.

_____

1.2 Sediments and Rocks!

(Rock/Sediment Samples in Class)   15 points

1.2 Sediments and Rocks!
(Rock/Sediment Samples in Class)   15 points

Answer Sheet   Name _____

Did you work with anyone? If so, write their names here as acknowledgement:
_____

Station 1: 1. _____   2. _____   3. _____

4. _____   5. _____

6. _____

7. _____

Station 2. 1. _____ 2. _____ 3. _____ 4. _____

5. _____   _____

6. _____

_____

Station 3. 1. _____   2. _____

Station 4. 1. _____

2. _____ 3. _____

4. _____

Station 5. 1. _____ 2. _____

3. _____   4. _____

Station 6. 1. _____   2. _____

3. _____

Station 7. 1. _____   2. _____   3. _____   4. _____

5. _____   6. _____

Station 8. 1. _____ 2. _____

3. _____

15

1.3 FEMA and the Government's Role in Disasters

(Internet Access, Graph Paper)   10 points

Name _____

Date _____

When people hear about a community or state that has been impacted by a flood, earthquake, or storm, they think of FEMA. FEMA is the Federal Emergency Management Agency and it is the agency responsible for the government's response to any disaster or significant event. In this assignment, you will visit various pages at FEMA's website. This site discusses how FEMA helps individuals in a disaster.

1. Go to http://www.fema.gov/rrr/inassist.shtm. What is the dominant form of assistance to individuals that FEMA gives?

2. What is the absolute maximum amount of money you can be given by FEMA (as an individual)? Would this cover the value of your house if it collapsed in an earthquake?

3. Navigate to http://www.fema.gov/library/dis_graph.shtm. (**Note**: there is an underscore in this URL: dis_graph.) On a separate piece of paper, graph the number of major disasters per year from 1953 onward. By the way, if you can use Excel, please feel free to copy and paste the data and have the computer do all the work.

4. Describe what you notice about this graph and its implications.

5. Go to http://www.fema.gov website. What are the currently active disasters in the Pacific Northwest (Oregon, Washington, Northern California, Idaho)? List the location and cause(s).

1.3 FEMA and the Government's Role in Disasters

(Internet Access, Graph Paper)   10 points

6. Look at the list of major disasters (just the major ones – no emergencies or fires) for the last three (3) years included in the table at http://www.fema.gov/library/dis_graph.shtm. (**Note:** there is an underscore in this URL: dis_graph.) You will need to follow the link from each year on the table to the list of major disasters for that year. How many of the total major disasters occurred in California? What percentage of the total number of disasters is this?

7. How many disasters occurred in the last three (3) years in West Virginia? Florida?

8. How many of the major disasters in the United States over the last three (3) years were hurricanes or tropical storms?

9. How many major disasters in the last three (3) years were earthquakes?

10. How many disasters in the last three (3) years were in Washington State? For each in our state, list the date, and what caused it.

11. Pick one of the Washington State disasters listed and tell me what happened in this disaster and how much aid was dispersed.

1.4 Introduction to Library Resources

(School Library, Word Processor)   10 points

Name _____

Due Date _____

This assignment involves a visit to the college library to answer the following questions. You will also be asked to write correctly formatted and complete bibliographic references for the journals, books, and articles you are asked to access. **Follow the examples in section <u>6.1 Reference Style Guide for Paper/Assignments</u> of this workbook.**

1.   What do you need to check out a book?

2.   Go to the periodicals area. What is the magazine that is shelved next to <u>Washington Geology</u>?

3.   Find the book *Volcanoes of North America* by Wood and Kienle. Look up Washington State and record the names of all the volcanoes listed for Washington State.

4.   Write a complete and correctly formatted bibliographic reference for *Volcanoes of North America*. (Note: This means a bibliographic reference for this book only. Please do not copy the bibliography of *Volcanoes of North America*!)

5.   Find the book *Puget Lowland Earthquakes of 1949 and 1965: reproductions of selected articles describing damage*, compiled by Gerald W. Thorsen. There are two copies of this book in the library, one on reserve and one that you can check out from the stacks (if no one has beat you to it). Write a complete and correctly formatted bibliographic reference for this book.

1.4 Introduction to Library Resources

(School Library, Word Processor)   10 points

6. The book is a collection of papers on the damage caused by the 1949 and 1965 earthquakes. There are descriptions of the damage for almost all Western Washington cities for these two quakes. Please find the description for your city/town and, *on a separate sheet of paper*, write the description for each of these two quakes. If there is no description for your city, use the closest city you can find.

7. Find the Circulation Desk in the library. This is where you might have to go to find videos that are important for this class. Find out the name of the friendly person working there and record it here.

8. Use ProQuest or a helpful reference librarian to assist you in finding a local newspaper article on one of the following: local earthquake hazards, local volcanic hazards, or local landslide events. These must be local in nature – i.e. not about an earthquake in another country or a volcano in Oregon. Ideally, it would relate to the city you live in. It should not be more than 1 year old. Write a complete and correctly formatted bibliographic reference for the article. You may want to print off the article for future use.

9. Find 3 OTHER books, maps or articles that you think might be helpful to your research paper. Write complete and correctly formatted bibliographic references for these three here. Please note: web pages are not acceptable.

10. Please type (in a word processing program) your answers to questions 4, 5, 6, 8 and 9. Turn this in with your assignment, but be sure to save the information. It will become part of your final paper – you've already started to work on it!!

1.5 Video: Plate Dynamics

10 points

Name _____

Due Date _____

(From the video "Plate Dynamics" - The Earth Revealed Series, program 6; available for review from the Circulation Desk in the school library)

1.   What is the name of layer of the Earth that the plates are made of?

2.   At what type of plate boundary is Iceland located?

3.   What is an example of a continent-continent divergent plate boundary?

4.   How and where is oceanic crust generated?

5.   Where are oceanic plates consumed or destroyed?

6.   How many types of convergent plate boundaries there are?

7.   What is the composition of magma erupted at ocean-ocean convergent plate boundaries?

8.   What is the composition of magma erupted at ocean-continent convergent plate boundaries?

9.   How do we get a continent-continent convergent plate boundary?

10. Which one is denser: oceanic or continental crust?

1.5 Video: Plate Dynamics

10 points

11. How deep can earthquake foci be at subduction zones?

12. Where are earthquakes more powerful: divergent or convergent plate boundaries? Why?

13. Do continent-continent convergent plate boundaries have active volcanoes?

14. What drives plate movement?

15. How is the lithosphere different from the asthenosphere?

16. What happens if you deform cold rocks rapidly?

17. What happens if you deform warm rocks slowly?

18. What are the sources of heat that cause convection?

19. In which other layer of the Earth (besides the mantle)is there convection?

20. What are mantle plumes?

1.5 Video: Plate Dynamics

10 points

21. How do we explain the fact that the Hawaiian Islands are made of rocks increasingly old as you move from SE to NW?

22. How long will it take until the island of Hawaii will be too far from the hot spot to be volcanically active?

23. How long will it take to Loihi to emerge above sea level?

24. What is a North American example of a continental hot spot?

1.6 Maps – Where Do You Live and What Do You Live On?

(Internet access, printer)   15 points

Name _____

Due Date _____

PART 1 – Where Do You Live? *HOMEWORK* (5 points)

Please complete this section before coming to class. Also note that if you do not complete this homework, you will find it much more difficult to complete the in-class assignment.

1. Go to www.mapquest.com

2. Type in your address under the "maps" section and click "Get Map"

3. Look on the right side of the map. The map that comes up will usually be zoomed to the second bar below the "zoom in" button which has a circle with a plus sign in it. When your mouse goes over it, it should say "zoom level 9." If you are on a different zoom level, click that second button.

4. Press ctrl+p <u>or</u> go to the top of the web browser under the file menu and select "print". **Do NOT use the print option within the web page (next to the word "maps")**

5. After successfully printing this map, click on the fourth bar down from the "zoom in" button which has a circle with a plus sign in it ("zoom level 7"). This will zoom out and show a much larger field of view.

6. Select "print" from next to the word "maps" <u>*on the web page*</u>. (Yes, this is exactly what you were told not to do in #4)

7. Click "send to printer"

8. Collect your two maps and bring them to class. Pay special attention to the location of nearby freeways, major streets, rivers, lakes, and the Puget Sound coastline as they show up on your Mapquest maps. Orienting yourself to these landmarks will assist you in locating your home on the soil and geologic maps.

PART 2 – What Do You Live On? *IN CLASS* (10 points)

(**NOTE**: Soil maps are also available from your instructor to be used in class or online at: http://www.or.nrcs.usda.gov/pnw_soil/wa_reports.html. Please note the two underscores in this link: pnw_soil and wa_reports. The online GIS system has some King County and all Pierce Country Soil Surveys. You must select the county you live in. The online system may be difficult to operator at first, but detailed instructions can be found by clicking on the "Need Help with Soil Maps Online?" at the top of the page.)

9.   From the map, what is the abbreviation for the type of soil that your house sits on?

1.6 Maps – Where Do You Live and What Do You Live On?
(Internet access, printer)   15 points

10.   From the key or legend, find out what this abbreviation stands for. What is the name of your soil type?

11.   From the non-technical descriptions, what type of material did your soil form on? (Example: material deposited by glaciers, water, glacial lake deposits, etc.)

12.   From the non-technical description, is your soil well drained? Does water move readily through it?

13.   From the non-technical description, what is the range of slopes this soil is found on? (Example: 0 to 6% slopes, 15 to 30% slopes, etc.)

14.   From the Water Features Table (K1), what is the flooding frequency for the soil that your house sits on?

15.   Write the complete bibliography in correct format for the soil map.

1.6 Maps – Where Do You Live and What Do You Live On?

(Internet access, printer)   15 points

Look at the local geologic map for the area you live in. If it is not available, use the King County map. These maps are also available online at http://geomapnw.ess.washington.edu/index.php?toc=maintoc&body=services/maps.htm. What is the abbreviation for the geologic unit (the stuff under the soil) that you live on?

16.   Write the name of the geologic unit you live on and the description of that unit.

17.   Your instructor will tell you whether to turn in this or take it home. If you are told to take it home, please type your answers to questions 9-17, **_and save them_**. You will use this in your end-of tem paper! Then bring the typed answers back on the assigned due date.

1.7 Plate Tectonic Setting of the Pacific Northwest
(Word processor, textbook/internet)   5 points

Name _____

Due Date _____

In the Pacific Northwest, we live at an active plate tectonic boundary. This tectonic setting is responsible for earthquake, tsunami, and volcanic hazards, and is a large contributing factor to landslide hazards in the region. In this activity, you will conduct a search of your text book, the library, and/or the internet to gather information about our place tectonic setting.

> **NOTE: The information you obtain for this assignment must be incorporated into the introduction of your Hazard and Risk Research paper.**

Instructions

Type a paragraph (or series of paragraphs) that describes the plate tectonic setting of the Pacific Northwest. Your paragraph should clearly indicate what type of plate boundary you are living near, what two plates are involved and where they are. You should also discuss any other types of plate tectonic boundaries in our vicinity (within a few hundred kilometers). What physical features of our region are the result of our plate tectonic setting?

A diagram or drawing may be helpful; please feel free to include this in your submission. You will need to clearly refer to the diagram in the text of your submission and have a caption for your diagram.

Your response should be typed and written in a clear, grammatically correct format. Your paragraphs should be well structured and smooth – with a topic sentence, for instance. You will be graded on the accuracy, completeness, and level of detail in your response, as well as the quality of your writing. Of course, if you use sources other than your class notes and text, you must include complete, correctly formatted citations within the text and a bibliography.

1.8 Support for You!
(Leg work)   5 points

Name _____

Due Date _____

1.   Please go to the Tutoring Center.

      What is the name of the tutor(s) for this class?

      What are their hours?

      Find a writing tutor. Get them to sign here: _____

      What can a writing tutor do to help you on your big paper?

2.   Go to my office. If I am there, get my initials: _____ If I am not
      there, write the time and date you were there, and write the number of
      geologic maps that are in the window.

3.   What are my office hours?

1.9 What is Plagiarism?

5 points

Name _____

Due Date _____

Plagiarism, or taking others ideas or words without giving credit, is a serious offense and will result in harsh penalties. If you have questions about this topic, please ask your instructor.

The following is the original (source) text:

"The seismology lab at the University of Washington records roughly 1,000 earthquakes per year in Washington and Oregon. Between one and two dozen of these cause enough ground shaking to be felt by residents. Most are in the Puget Sound region, and few cause any damage. However, based on the history of past damaging earthquakes and our understanding of the geologic history of the Pacific Northwest, we are certain that damaging earthquakes (magnitude 6 or greater) will recur in our area, although we have no way to predict whether this is more likely to be today or years from now." (PNSN, 2002)

1. Would the following be considered plagiarism?

> There are ~ 1,000 earthquakes per year in Washington. 12-24 of these cause enough ground shaking to be felt by residents.

> Why or why not?

2. Would the following be considered plagiarism?

> There are ~ 1,000 earthquakes per year in Washington. 12-24 of these cause enough ground shaking to be felt by residents. (PNSN, 2002)

> Why or why not?

3. Would the following be considered plagiarism?

> According to the Pacific Northwest Seismograph Network, there are "roughly 1,000 earthquakes per year in Washington and Oregon. Between one and two dozen of these cause enough ground shaking to be felt by residents." (PNSN, 2002)

> Why or why not?

4. Would the following be considered plagiarism?

> There are about one thousand earthquakes every year in The Pacific Northwest. Some of these can be felt. Most are in the urbanized areas around Seattle, but few cause destruction.

> Why or why not?

1.9 What is Plagiarism?

5 points

5. Would the following be considered plagiarism?

There are about one thousand earthquakes every year in The Pacific Northwest.   Some of these can be felt. Most are in the urbanized areas around Seattle, but few cause destruction. (PNSN, 2002)

Why or why not?

6. Would the following be considered plagiarism?

You work with someone on an assignment. You do half of the problems and they do half of the problems. You then share the answers and submit your own paper with all the answers on it.

Why or why not?

7. Would the following be considered plagiarism?

You work with someone on a homework assignment. You both work on each problem, as a team. You then each submit the individually homework under you own name.

Why or why not?

8. Would the following be considered plagiarism?

You work with someone on a homework assignment. You both work on each problem, as a team. You then each submit the homework individually under you own name, however you note that you worked with your classmate.

Why or why not?

The correctly formatted, complete bibliography for the source text is:

Pacific Northwest Seismic Network, 2002. Earthquake Hazards in Washington and Oregon. <http://www.pnsn.org/INFO_GENERAL/eqhazards.html> Accessed September 19, 2005.

2.1 Identifying Faults

10 points

Name _____
Due Date _____

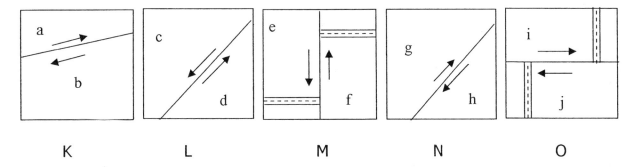

K　　　　　L　　　　　M　　　　　N　　　　　O

Refer to the above diagrams. Diagrams K, L, and N are side views (cross sectional views) M and O are top-down views (map views) with an off-set road. Fill in the appropriate letter(s).

_____ are hanging walls

_____ are footwalls

_____ are/is caused by shearing

_____ are/is caused by compression

_____ are/is caused by tension

_____ are/is a normal fault

_____ are/is a reverse fault

_____ are/is a thrust fault

_____ are/is a right-lateral strike-slip fault

_____ are/is a left-lateral strike slip fault

_____ would be more common at a convergent plate boundary

_____ would be more common at a divergent plate boundary

_____ would be more common at a transform plate boundary

35

2.1 Identifying Faults

10 points

Draw the continuation of the bed to show a normal fault on the left and a reverse fault on the right.

Normal   **Reverse**

   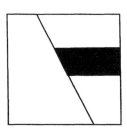

36

2.2 Earthquake Information on the Internet

(Internet Access)   5 points

Name _____

Due Date _____

The USGS Earthquake Hazards Program maintains the National Earthquake Information Center database at the following website address:

http://neic.usgs.gov/current_maps.html
There is an underscore in this URL: current_maps

1. Go to the website. Note the date and time that the maps have been updated:

2. Follow the link to the map of the Pacific Northwest (United States). How many earthquakes occurred in Washington in the past 2 days?

3. Have any occurred within the last 2 hours (at the time you checked)?

4. Notice the epicenters of the earthquakes are marked by squares. What does the size of the squares mean?

5. Click on the one of the epicenter squares. At the bottom of the map, some information is given. Answer the following questions:

What is the date and time of the earthquake you chose?

What was the magnitude of the earthquake?

What was the location of the earthquake?

Did you feel this earthquake? If not, why do you think you didn't?

2.3 Epicenters and Magnitude
(Ruler, compass, pencil)   10 points

Introduction

As plates move, **stress** builds up at their margins, like a stick that you try to break across your knee. Eventually, the stress overcomes the strength of the rocks and they break and shift along **faults**. The shifting releases elastic energy that had accumulated, in the form of **seismic waves**. Seismic waves move outward in all directions from the **focus** of the earthquake, the exact place where the breaking and shifting of rocks begins. When seismic waves reach the Earth's surface they cause the shaking of the ground that people feel – an **earthquake**.

A **seismograph** is a machine that can detect and record on paper the shaking produced by seismic waves. This result in a graph made of a series of characteristic squiggles called a **seismogram**. Careful analysis of seismograms shows that there are several different types of waves produced in an earthquake. The first two to be recorded (and the ones we will use in this lab) are:

> **P-waves** – Primary or pressure waves. These are the fastest traveling waves and the first recorded on a seismogram. People often report that they feel like a large truck going by, a distant rumble or even a blast. They do not cause much shaking.

> **S-waves** – Secondary or shear waves. These are slower than P waves and are the second waves to arrive at the seismograph. They cause a lot of shaking and often do most of the damage in an earthquake.

You will be analyzing a seismogram to determine some basic information about the earthquake that produced it, including the location of the epicenter and Richter Magnitude. But before you get started, let's use an example.

2.3 Epicenters and Magnitude
(Ruler, compass, pencil)

10 points

Instructions and Example

An example of a seismogram recorded at a seismograph in Des Moines, WA is shown in Figure 1. Scientists use the first significant displacement from the horizontal line to determine the arrival time. From this seismogram you can see that the P-wave arrived at 8:00:12 (12 seconds after 8 a.m.), and the S waves arrived at 8:00:24.

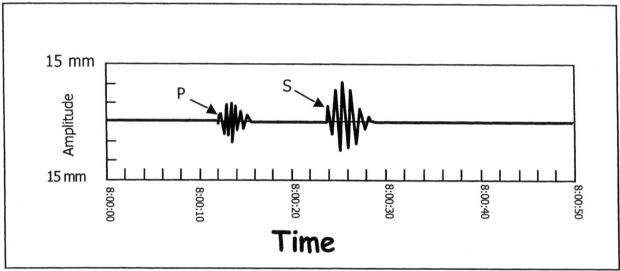

Figure 1. Seismogram for earthquake recorded in Des Moines.

What the seismogram does not indicate is where the epicenter was and when the earthquake occurred. (The P-wave arrived in Des Moines at 8:00:12, but it started sometime before, when the earthquake occurred.)

Distance from Epicenter

To discover how far away the earthquake epicenter was, and when the earthquake occurred, you will need to use the time-travel graph shown in Figure 2. This graph shows the time required for the seismic waves to travel a particular distance. Because P-waves travel faster than S-waves, as the waves travel away from the epicenter, the time-lag between the two waves increases. You can use this time lag to determine the distance from the epicenter.

In order to determine the distance to the epicenter, you will need to determine the time-lag between the P-and S-wave arrival times, to do this:

1. Determine the arrival time of the P and S waves.
2. Subtract the P-wave arrival time from the S-wave arrival time.
3. Place a piece of paper along the time axis of the time travel graph (Figure 2) and mark off the length corresponding to the time-lag you just determined

2.3 Epicenters and Magnitude
(Ruler, compass, pencil)   10 points

4. Slide the piece of paper along the P and S-wave time travel curves until the two marks line up. Be sure to keep the paper vertical!

5. Where they line up, read straight down to find the distance to the epicenter.

6. Practice on the Des Moines seismogram (Figure 1). You should get a time-lag of 12-seconds, which corresponds to a distance of about 100 km.

Location of Epicenter
To find the location of the epicenter, three seismographs are needed. Each one determines the distance from the epicenter. On a map you draw three circles around the locations of the seismographs with radii that are the appropriate distance. There should be one place where the circles intersect, the epicenter of the quake. In some cases small errors make the circles not intersect exactly, but a point closest to the three is a good estimation.

Origin Time
Once you know how far away the earthquake was, you can find the time the earthquake occurred. By looking at Figure 2 you can determine how long it takes a S-wave to travel any distance. Looking at the S-wave curve, you can read on the time axis that it will take a S-wave 24 seconds to travel the 100 km from the epicenter. Because the S-wave arrived 24 seconds after 8 a.m., the earthquake must have happened at 8:00:00.

The Magnitude of the Earthquake
Once the distance to the earthquake is found, the amplitude of the seismic waves can be used to determine the Richter magnitude of the earthquake. A diagram called a **nomogram** (Figure 5) correlates the distance, amplitude and the magnitude of the earthquake. To determine the magnitude follow these steps:

1. On the left line of figure 5 find the distance to the epicenter from the seismogram you were using.

2. Measure the amplitude of the S-waves from the seismogram. The amplitude is the total vertical displacement from the center line to the lowest point or the highest point of the seismogram's "squiggle."

3. On the right side of the nomogram, make a mark at the correct amplitude.

4. Connect the distance with the amplitude with a line. It will pass through the middle line. Where the line crosses the middle scale is the Richter magnitude.

41

2.3 Epicenters and Magnitude
(Ruler, compass, pencil)   10 points

Figure 2. Time-travel graph for P and S waves.

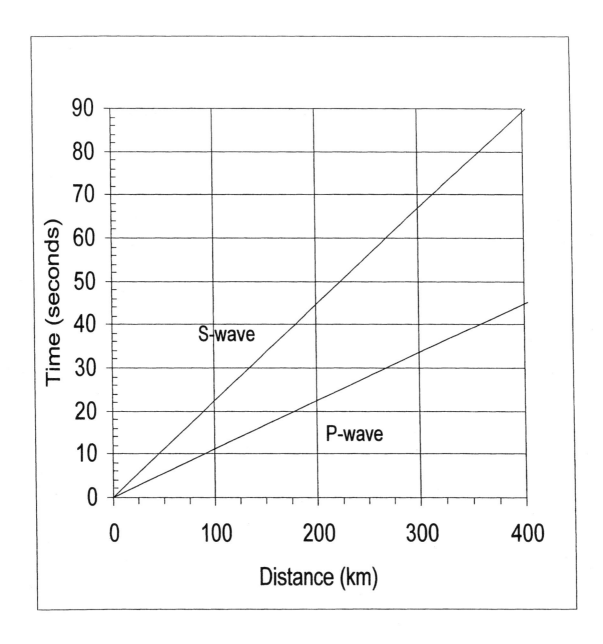

2.3 Epicenters and Magnitude
(Ruler, compass, pencil)   10 points

Name _____

Date Due _____

Practice questions for this exercise

1. Read the introduction to this exercise

2. Using Figure 2, answer the following question:

   An earthquake occurs 200 kilometers from a seismograph. If the earthquake occurred at 7:13:56, what time will the P and S waves arrive at the seismograph?

   P Wave:_____

   S-Wave:_____

3. Look at the seismogram from Des Moines in Figure 1. How far is this station from the epicenter?

4. What is the magnitude of the earthquake felt in Des Moines that is shown in Figure 1? Use the nomogram (Figure 5) to answer this question.

2.3 Epicenters and Magnitude
(Ruler, compass, pencil) 10 points

Figure 3. Seismograms for an Earthquake in the Pacific Northwest.

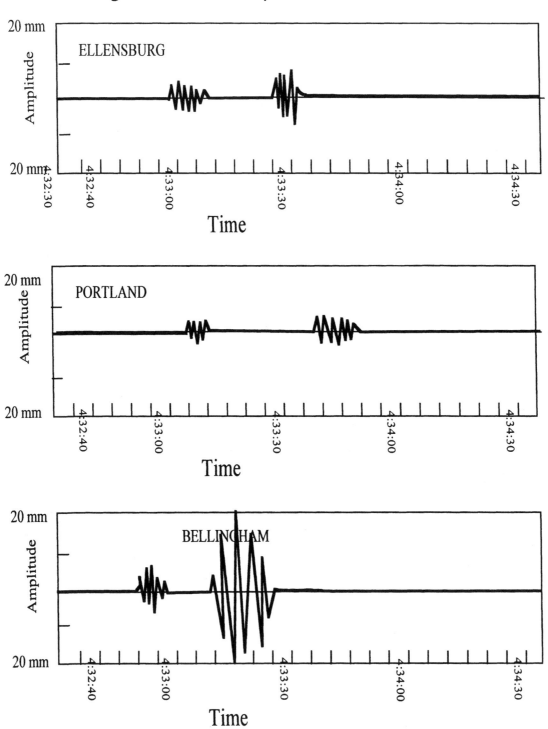

2.3 Epicenters and Magnitude

(Ruler, compass, pencil)   10 points

Exercise Questions

Complete the questions on this page and submit them with Figure 4 by the due date given by your instructor.

1. Look at the three seismograms in Figure 3, and fill in the table for each seismic station.

	P-wave arrival time	S-wave arrival time	Lag time between P and S-waves	Amplitude
Ellensburg				
Portland				
Bellingham				

2. Using the time-travel graph determine the distance from each seismograph to the epicenter.

   Ellensburg _____km

   Portland   _____km

   Bellingham _____km

3. Using a drafting compass, determine the location of the epicenter on the map of Washington (Figure 4).

4. Origin Time of the earthquake:

   What is the arrival time of the S-wave in Portland?

   _____

   How far is Portland from the earthquake epicenter?   _____

   How long did it take the S-wave to travel from the epicenter to Portland?

   _____

   At what time did the earthquake occur?   _____

45

2.3 Epicenters and Magnitude
(Ruler, compass, pencil)   10 points

5. What was the Richter Magnitude of the earthquake? Use the amplitude from the nearest station.

_____

If the shaking was 10 times worse, what would the Richter Magnitude be?

2.3 Epicenters and Magnitude
(Ruler, compass, pencil)   10 points

Figure 4. Map of seismograph locations.

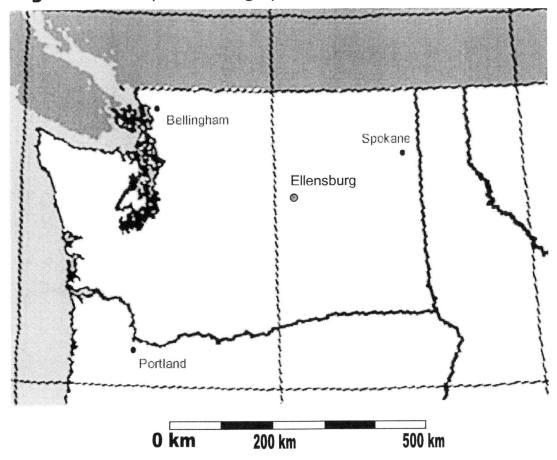

Figure 5. Richter nomogram for determining magnitude of an earthquake.

2.4 Calculating Earthquake Magnitude
(Calculator)   10 points

___

Name _____

Due Date _____

Please show your work or explain your reasoning.

1. In the 1964 Alaska earthquake (Magnitude 8.3), a seismogram at University of Alaska - Fairbanks showed squiggles of amplitude 100 centimeters. What would be the amplitude of the squiggles at this same seismograph in a Magnitude 6.3 aftershock?

2. The 6.3 aftershock of the Alaska earthquake released about the same amount of energy as 500 atomic bombs. What was the equivalent energy released by the main quake (in terms of atomic bombs)?

3. How much energy (in terms of atomic bombs again) was released by the magnitude 7.3 earthquake that occurred in Loma Prieta, CA in 1989?

2.5 Seismic Design
(Poster board & tape supplied in class)   10 points
Original Author: Larry Braille, Purdue University

Name _____

Group Member _____

Group Member _____

Group Member _____

Date _____

Instructions

You have 40 minutes to make a building at least 30 cm and 3 stories high. Your goal is to make the most seismically resistant design with the materials provided.
You have been given:

(4)   8x8 cm squares (floors and roofs)

(12)   1.5x10 cm (uprights)

(12)   1.5x15 cm strips (reinforcing)

(1)   32 x 10 cm to use and cut as you wish.
        1 m of Scotch tape

You will then test your building for shaking as explained by the instructor.

1.   Describe any techniques you used to make your building stronger.

2.   Attach the accelerometers and shake the building. What was the difference in acceleration between the base and the top of your building?

3.   Did this vary with the frequency of shaking?

2.5 Seismic Design
(Poster board and tape supplied in class)   15 points

4. Tape one or more weights onto your building. How much weight did you building hold?

5. Again attach the accelerometers and shake the building. Discuss the failure of your building. What failed? Which floor? What could have been done to stop this failure?

6. What techniques did others use that you wished you had used?

From your notes or your book, answer the following questions:

7. What is a soft story?

8. What is a shear wall? How do shear walls act to prevent or lessen earthquake damage?

9. What is a cross brace? How does a cross brass act to prevent or lessen earthquake damage?

2.6 Shaking during the 2001 Nisqually Earthquake

(Colored Pencils)   15 points

Name _____

Due Date _____

On February 28, 2001, a magnitude 6.8 earthquake struck the Puget Sound region and caused nearly $2 billion in damage. In part 1 of this assignment, you will use real damage reports from several locations to determine the Modified Mercalli Intensity of shaking. You will then create a seismic intensity map for this quake. In part 2 of this assignment, you will use measurements of shaking determined by seismograph recordings to look at the relationship between shaking and damage.

References:

Dewey, et al., 2002, *Intensity Distribution and Isoseismal Maps for the Nisqually, Washington, Earthquake of 28 February 2001,* USGS Open File Report 02-346. (excepts of shaking descriptions)

Foley, D. et al. Investigations in Environmental Geology 2/e, Prentice Hall, 1999 Exercise II-3 was the original idea for this type of exercise.

Part 1: Assigning Mercalli Intensities/Creating an Intensity Map

The **intensity** of an earthquake at a site is based on the observations of individuals during and after the earthquake. It represents the severity of the shaking as perceived by those who experienced it. It is also based on observations of damage to structures, movement of furniture, and changes in the Earth's surface as a result of an earthquake.

The **Modified Mercalli Intensity Scale (MMI)** is commonly used to quantify intensity descriptions. The scale uses roman numerals ranging from I to XII. The scale is shown in Figure 1.

A **Seismic Intensity Map** shows the distribution of seismic intensities associated with an earthquake. The greatest impact of an earthquake is usually in the region of the epicenter, with lower intensities occurring in nearly concentric zones outward from this region. The quality of construction and variation of geological conditions affect the distribution of intensity. In this exercise we will use historical descriptions from the 2001 Nisqually earthquake in western Washington to determine Modified Mercalli Intensities. A map of Washington and a data grid are included for you to record your work.

2.6 Shaking during the 2001 Nisqually Earthquake

(Colored Pencils)   15 points

Instructions:

1. Divide into groups and read the descriptions by people who experienced the earthquake in a variety of locations.

2. Compare the descriptions to the Modified Mercalli Scale and assign an intensity value. When selecting an intensity value always **pick the lowest intensity** value that can explain all the damage or characteristics given in the descriptions.

3. Record the intensity value in the table provided.

4. Place the intensity values from the table on the map next to the appropriate location.

5. Contour the data on the map by drawing lines to separate the different intensities. Be sure to indicate the outer limits of the area where the earthquake was experienced. Your contour maps should resemble a bull's eye pattern with concentric loops of each intensity. The highest earthquake intensity will be in the center. You should color in the entire area between lines, once you are certain your lines are drawn in the right place. Use the following colors:

I-III: Dark Blue
IV: Light Blue
V: Green
VI: Yellow
VII: Orange
VIII: Red
IX -XI: Dark Red (maroon/brick red)

Figure 1. Modified Mercalli Intensity Scale of 1931.

I	Not felt by people, except under especially favorable circumstances. However, dizziness or nausea may be experienced. Sometimes birds and animals are uneasy or disturbed. Trees, liquids, bodies of water may sway gently.
II	Felt indoors by a few people at rest, especially on upper floors of multi-story buildings. As in Grade I, birds and animals are disturbed, and trees, structures, liquids, and bodies of water may sway. Delicately suspended hanging objects may swing.
III	Felt quite noticeably indoors; especially on upper floors of buildings. Hanging objects swing. Vibration like passing of light trucks. May not be recognized as an earthquake. Standing automobiles may rock slightly. Duration may be estimated.
IV	Hanging objects swing. Vibration like passing of heavy trucks; or a jolt like a heavy ball striking the walls. Standing automobiles rock noticeably. Windows, dishes, doors rattle. Glasses clink. Wood walls and frames creak. Felt indoors by many during the day; outdoors by few. Some awakened at night. Liquids in open vessels are disturbed slightly.
V	Felt outdoors; direction estimated. Sleepers awakened. Liquids disturbed, some spilled. Small unstable objects displaced or upset. Doors swing, close, open. Shutters, pictures move. Pendulum clocks stop, start, change rate. Some windows may be broken; cracked plaster in a few places. Disturbances of trees, poles, and other tall objects sometimes noticed (slight shaking).
VI	Felt by all. Many frightened and run outdoors. Persons walk unsteadily. Windows, dishes, glassware broken. Knickknacks, books, etc. off shelves. Pictures off walls. Furniture moved or overturned. Weak plaster and masonry D cracked. Small bells ring. Trees, bushes shaken (visibly or heart to rustle). Damage slight. Liquids are set in strong motion.
VII	Frightens everyone. Difficult to stand. Noticed by drivers of automobiles. Hanging objects quiver. Furniture broken or overturned. Damage to masonry D, including cracks. Weak chimneys broken at roofline. Fall of plaster, loose bricks, stones, tiles cornices, parapets. Some cracks in masonry C. Waves on ponds; water turbid with mud. Small slides and caving in along sand or gravel banks. Large bells ring. Damage is negligible in building of good design and construction; slight to moderate in well-built ordinary buildings; considerable in poorly built structures.
VIII	Alarm approaches panic. Steering of moving cars affected. Trees shake strongly, and branches break off. Sand and mud erupt in small amounts. Changes in flow of wells and springs. Decays pilings broken off. Cracks in wet ground and on steep slopes. Heavy furniture overturned. Damage slight in specially designed structures; considerable in ordinary substantial buildings with partial collapse; great in poorly built structures. Twisting, fall of chimneys, factory stacks, towers, elevated tanks. Frame houses moved on foundations if not bolted down; loose panel walls thrown out.
IX	General panic. Ground cracks conspicuously. Damage considerably in specially designed structures; well-designed frame structures thrown out off plumb; great in substantial buildings, with partial collapse. Buildings shifted off foundations. Underground pipes broken. Sand boils develop. Serious damage to reservoirs.
X	Most masonry and frame structures and foundations are destroyed. Well-built wooden structures and bridges are severely damaged, and some collapse. Ground, especially where loose and wet, cracks up to widths of several inches; Sand and mud shift horizontally on beaches and flat land. Large landslides. Water is splashed (slopped) over banks. Open cracks and broad wavy folds open in cement pavements and asphalt road surfaces. Dams, dikes, embankments are seriously damaged. Railroad rails bend.
XI	Few (masonry) structures remain standing. Damage is severe to wood frame structures. Bridges destroyed. Broad fissures in ground. Underground pipelines completely out of service. Earth slumps and land slips in soft ground.
XII	Damage is total. Large rock masses displace. Waves seen on ground surfaces. Lines of sight and level distorted. Objects thrown upward into air

2.6 Shaking during the 2001 Nisqually Earthquake
(Colored Pencils)

15 points

DATA TABLE – Earthquake Descriptions

Mercalli Intensity (I-XII)	City Abbr.	Description
	AL	**Aldergrove B.C.** Most people felt the quake. Slight Shaking.
	AN	**Anacortes.** T he 145-foot tall Morrison Mill smokestack, built in 1926, was cracked and subsequently taken down (Seattle Post-Intelligencer, March 5, 2001). Percentages of effects noted by 29 internet respondents: difficulty standing or walking during the earthquake— 21%; objects toppled over or fell off shelves—17%; pictures on walls moved or were knocked askew—34%.
	AS	**Astoria OR.** Percentages of effects noted by 24 internet respondents: difficulty standing or walking during the earthquake—21%; objects toppled over or fell off shelves—38%; pictures on walls moved or were knocked askew—38%
	AT	**Anatone.** Not Felt
	BF	**Boeing Field** . An observer via Internet reported, "Nearly all bookcases and some file cabinets fell and the computer on my desk fell to the floor. Large cracks appeared in the stairwells, and lots of fine concrete dust filled the air. Linoleum on the first floor near stanchions was buckled and raised in about a one-foot radius around each stanchion that I saw." A nearby observer in an office on East Marginal Way noted, "When I first came out of the building there was water pouring out from the sidewalk."
	BR	**Bremerton.** Many Kitsap County schools experienced minor damage, including cracked walls and fallen ceiling tiles. Elevators were damaged at Harrison Hospital and plumbing of the heating system was damaged. A grandstand at the west end of Thunderbird Stadium was damaged (press reports). A number of homes were damaged in the area between 11th and Burwell Streets, and between Warren and Naval Avenues. The damage typically involved cracked or fallen chimneys. Many chimneys were damaged in West Bremerton (press reports). "Probably 30 to 50 [Bremerton homes] took major hits, with roofs buckling and walls collapsing" (Seattle Post-Intelligencer, March 3, 2001, p. A6, col. 4). The exterior walls of some commercial buildings were cracked and some windows were broken. Many items fell from shelves in food stores, and ceilings were damaged (press reports).
	BU	**Buckley/Wilkeson.** At the entrance to Wilkeson a 76-year-old arch consisting of two 20-foot-high columns of Wilkeson sandstone connected at the top by a large fir log was damaged by the earthquake and had to be removed (Tacoma News Tribune, March 1, 2001). The USGS field team found that bricks fell from a few old chimneys, but many old chimneys and the downtown's old unreinforced masonry buildings were undamaged. A few items fell from grocery shelves. Many people had difficulty walking. objects toppled over or fell off shelves—

DATA TABLE – Earthquake Descriptions

Mercalli Intensity (I-XII)	City Abbr.	Description
	CB	***Clallam Bay.*** Percentages of effects noted by two internet respondents: difficulty standing or walking during the earthquake—50%; objects toppled over or fell off shelves—50%; pictures on walls moved or were knocked askew—100%; furniture or appliances slid, toppled over, or became displaced—50%.
	CE	***Centralia.*** A cornice was damaged on a downtown hardware store and bricks fell from the tops of several old unreinforced masonry buildings. An abandoned one-story unreinforced brick building that had once housed a garage and auto-glass shop was damaged by the earthquake and collapsed seven days after the earthquake. A large plate-glass window was broken in an automobile dealership. At Centralia College about 35 ceiling tiles and 1 lighting fixture fell. In residential areas near Centralia College the USGS field team found approximately 10% of old chimneys were damaged. (The Chronicle, March 1, 2001, p. A10 and engineering reports and USGS field team). Many objects toppled over or fell off shelves and pictures on walls moved or were knocked askew. Few cracks in walls.
	CH	***Chelan.*** Percentages of effects noted by nine internet respondents: difficulty standing or walking during the earthquake—11%; objects toppled over or fell off shelves—33%; pictures on walls moved or were knocked askew—67%; furniture or appliances slid, toppled over, or became displaced—11%.
	CI	***Cinebar.*** Postal questionnaires reported a few old chimneys cracked; a few windows cracked. A few pictures on walls moved or were knocked askew, furniture or appliances slid. Small wall cracks noticed.
	CL	***Clatskanie OR.*** Roof beams shifted in the old gymnasium of Clatskanie Elementary School (press reports). The USGS field team observed isolated chimney damage. Light disturbance to stock was reported in a convenience store. Percentages of effects noted by three internet respondents: difficulty standing or walking during the earthquake— 67%; objects toppled over or fell off shelves—33%; pictures on walls moved or were knocked askew—100%; furniture or appliances slid, toppled over, or became displaced —33%; a few large cracks in walls—33%.
	CLE	***Cle Elum.*** Percentages of effects noted by 15 internet respondents: difficulty standing or walking during the earthquake—20%; objects toppled over or fell off shelves—60%; pictures on walls moved or were knocked askew—80%; furniture or appliances slid, toppled over, or became displaced—13%; hairline cracks in walls—27%
	CN	***Concrete.*** Postal questionnaires reported: exterior and interior walls sustained hairline cracks; a few old chimneys cracked; a few windows cracked. Percentages of effects noted by nine internet respondents: difficulty standing or walking during the earthquake— 33%; objects toppled over or fell off shelves—33%; pictures on walls moved or were knocked askew—56%; hairline cracks in walls—22%.

DATA TABLE – Earthquake Descriptions

Mercalli Intensity (I-XII)	City Abbr.	Description
	CO	***Coombs B.C.*** Cracks were reported in interior walls. Wind chimes sounded.
	COU	***Coupeville.*** Percentages of effects noted by 14 internet respondents: objects toppled over or fell off shelves— 29%; pictures on walls moved or were knocked askew— 50%
	CP	***Copalis Beach.*** Postal questionnaires reported: a few old chimneys cracked or lost bricks; a few windows cracked. Effects noted by one internet respondent: pictures on walls moved or were knocked askew; hairline cracks in walls; many large cracks in walls; ceiling tiles or lighting fixtures fell.
	CQ	***Coquitlam B.C.*** A few books fell off shelves and doors swung open and close.
	CV	***Colville.*** Creaking of houses noticed and some people felt it was like a truck impacting their building.
	DA	***Darrington.*** Percentages of effects noted by six internet respondents: difficulty standing or walking during the earthquake—17%; objects toppled over or fell off shelves—83%; pictures on walls moved or were knocked askew—67%; furniture or appliances slid, toppled over, or became displaced—33%; hairline cracks in walls—33%.
	DU	***DuPon.*** Books fell off shelves and artwork fell (Tacoma News Tribune, March 1, 2001). The USGS field team observed isolated chimney damage including a couple of old chimneys that lost a few bricks. About a third of respondents reported: difficulty standing or walking during the earthquake; many objects toppled over or fell off shelves and pictures on walls moved or were knocked askew. Furniture or appliances slid, toppled over, or became displaced. Many hairline cracks in walls with a few large cracks in wall.
	EL	***Elbe.*** Postal questionnaires reported several old chimneys cracked or lost bricks; a few windows cracked.
	EM	***Elma.*** The USGS field team noted isolated chimney damage and visited a grocery store where only a few things fell. Percentages of effects noted by 11 internet respondents: difficulty standing or walking during the earthquake—45%; objects toppled over or fell off shelves—73%; pictures on walls moved or were knocked askew—64%; furniture or appliances slid, toppled over, or became displaced—45%; hairline cracks in walls—36%; a few large cracks in walls—18%; ceiling tiles or lighting fixtures fell—18%.
	EU	***Eugene OR.*** Barely Felt

DATA TABLE – Earthquake Descriptions

Mercalli Intensity (I-XII)	City Abbr.	Description
	EV	**Everett (downtown).** Postal questionnaires reported: a few mobile homes fell off their foundations; exterior walls sustained large cracks; interior walls sustained a few large cracks and split at seams; a few old chimneys twisted, leaned, lost bricks or fell; some windows were broken out; in some buildings almost all small objects overturned and fell and almost all knickknacks broke; almost all items were shaken off store shelves; large furniture and heavy appliances were displaced; retaining walls partially fell; a few tombstones twisted or fell. Percentages of effects noted by 46 internet respondents: difficulty standing or walking during the earthquake— 35%; objects toppled over or fell off shelves—35%; pictures on walls moved or were knocked askew—46%; furniture or appliances slid, toppled over, or became displaced —17%; hairline cracks in walls—17%.
	FL	**Fort Lewis.** A parapet failed in a military dormitory (PEER, p. 21). An internet respondent in an office within a maintenance facility at Fort Lewis said, "The floor started to pitch up and down. The 12 ft X 12 ft room began to twist from left to right. We ran out the open bay door. In the parking lot, 2.5-ton trucks and five Hummers pitched back and forth. The parking lot looked as if there were waves rolling through the ground. While walking I felt as though I suddenly became heavier and then lighter."
	GH	**Gig Harbor.** In some stores many items fell from shelves, and furniture overturned (Tacoma News Tribune, March 1, 2001). Percentages of effects noted by 10 internet respondents: difficulty standing or walking during the earthquake— 50%; objects toppled over or fell off shelves—80%; pictures on walls moved or were knocked askew—70%; furniture or appliances slid, toppled over, or became displaced —30%; hairline cracks in walls—20%; cracks in chimney—10%.
	GL	**Glenoma.** Postal questionnaires reported: several old chimneys lost bricks and a few twisted, leaned or fell; many small objects overturned and fell and knickknacks broke; large furniture and heavy appliances were displaced; many items were shaken off store shelves.
	GR	**Graham.** Percentages of effects noted by 24 internet respondents: difficulty standing or walking during the earthquake—46%; objects toppled over or fell off shelves—88%; pictures on walls moved or were knocked askew—79%; furniture or appliances slid, toppled over, or became displaced—42%; hairline cracks in walls—4%. car.' He was not scared." Percentages of effects noted by 67 internet respondents: difficulty standing or walking during the earthquake— 45%; objects toppled over or fell off shelves—79%; pictures on walls moved or were knocked askew—75%; furniture or appliances slid, toppled over, or became displaced
	HO	**Hope,B.C.** Most people felt it, but no damage

DATA TABLE – Earthquake Descriptions

Mercalli Intensity (I-XII)	City Abbr.	Description
	IS	**Issaquah.** An internet respondent said, "I experienced violent shaking and jerking. Computer monitors and large heavy bookshelves fell over. Water pipes burst. Glass in the entry shattered." Percentages of effects noted by 61 internet respondents: difficulty standing or walking during the earthquake— 54%; objects toppled over or fell off shelves—79%; pictures on walls moved or were knocked askew—67%; furniture or appliances slid, toppled over, or became displaced —49%; hairline cracks in walls—28%; a few large cracks in walls—13%
	KE	**Kent (Midway).** Near Second Ave. and Harrison St. an unreinforced masonry wall of a single-story warehouse partially collapsed. Masonry fell from the front of a building on Second Ave. South (Tacoma News Tribune, March 1, 2001, p. A10. col. 6). Windows shattered in a department store on Meeker St. (Tacoma News Tribune, March 1, 2001, p. A10. col. 6). Percentages of effects noted by 54 internet respondents: difficulty standing or walking during the earthquake—63%; objects toppled over or fell off shelves—81%; pictures on walls moved or were knocked askew—61%; furniture or appliances slid, toppled over, or became displaced—39%; hairline cracks in walls—30%; a few large cracks in walls—13%; many ceiling tiles or lighting fixtures fell—28%;
	KL	**Kelso.** Postal questionnaires reported: drywall split at seams; in some buildings almost all small objects overturned and fell; heavy appliances were displaced by inches; many items were shaken off store shelves. Percentages of effects noted by 24 internet respondents: difficulty standing or walking during the earthquake— 42%; objects toppled over or fell off shelves—50%; pictures on walls moved or were knocked askew—67%; furniture or appliances slid, toppled over, or became displaced—17%; hairline cracks in walls—33%.
	LA	**Lacey.** A woman had just brought her 2-year-old son into her mother's kitchen in Lacey when the shaking started. "Stuff all around us was crashing. Dishes were falling off the open shelves, water was splashing out of the pot of potatoes on the stove and one potato piece actually flew out of the pot. A flying cup hit me on the wrist. Three 19-inch TVs fell in three different rooms, ceramic figurines were broken in the cabinets, cupboards and drawers were opened and some dishes were broken throughout the house. The dog's water dish (on the basement floor) completely emptied. A concrete 2-foot statue outside tipped over. My 3 1/2-year-old was still in the car in his car seat. He later told me that he thought 'Grandma's dog shake car.' He was not scared." Percentages of effects noted by 67 internet respondents: difficulty standing or walking during the earthquake— 45%; objects toppled over or fell off shelves—79%; pictures on walls moved or were knocked askew—75%; furniture or appliances slid, toppled over, or became displaced —36%; hairline cracks in walls—28%; a few large
	LE	**Lewiston ID.** Barely Felt

DATA TABLE – Earthquake Descriptions

Mercalli Intensity (I-XII)	City Abbr.	Description
	LO	**Longview.** Ceiling tiles fell at the Fred Meyer store (Daily News, March 1, 2001). The USGS field team visited a WalMart where enough items fell to fill a dozen shopping carts. Percentages of effects noted by 68 internet respondents: difficulty standing or walking during the earthquake— 40%; objects toppled over or fell off shelves—47%; pictures on walls moved or were knocked askew—54%; hairline cracks in walls—22%
	ML	**Moses Lake.** Felt like a truck went by Many did not know it was an earthquake.
	OL	**Olliver B C .** Telephone poles noticed to swing, Some people awakened.
	OM	**Omak.** Automobiles noticed to rock slightly. Rattling indoors.
	OP	**Ocean Park.** Percentages of effects noted by four internet respondents: difficulty standing or walking during the earthquake—100%;pictures on walls moved or were knocked askew—100%; A few hairline cracks in walls.
	PA	**Pacific.** Percentages of effects noted by six internet respondents: difficulty standing or walking during the earthquake—67%; objects toppled over or fell off shelves—83%; pictures on walls moved or were knocked askew—67%; furniture or appliances slid, toppled over, or became displaced—33%; hairline cracks in walls— 33%.
	PA	**Parksville B.C.** Objects rocked, some hanging pictures swung
	PC	**Packwood.** Percentages of effects noted by seven internet respondents: difficulty standing or walking during the earthquake—29%; objects toppled over or fell off shelves—57%; pictures on walls moved or were knocked askew—86%; furniture or appliances slid, toppled over, or became displaced—14%; hairline cracks in walls—14%; one or several windows cracked—14%; cracks in chimney —14%.
	PK	**Parksville B.C.** Chandeliers swung, Glass objects rattle.
	PO	**Port Angeles.** Percentages of effects noted by 50 internet respondents: difficulty standing or walking during the earthquake—26%; objects toppled over or fell off shelves—10%; pictures on walls moved or were knocked askew—38%; furniture or appliances slid, toppled over, or became displaced—16%; hairline cracks in walls—14
	PR	**Powell River BC.** *Barely felt*
	PU	**Puyallup.** The USGS field team found slight damage to walls of some unreinforced masonry buildings and a few fallen chimneys. A large supermarket experienced mild disturbance of stock, with some items off shelves in nearly every aisle; typical shelves in the store had raised edges that probably kept some items from sliding off. An observer commented, "It wasn't the shaking or noise that woke me up but rather the pictures that were falling on my head from the wall."

DATA TABLE – Earthquake Descriptions

Mercalli Intensity (I-XII)	City Abbr.	Description
	QU	**Quilcene.** Percentages of effects noted by four internet respondents: difficulty standing or walking during the earthquake—50%; objects toppled over or fell off shelves—100%; pictures on walls moved or were knocked askew—75%.
	RA	**Ravensdale.** Postal questionnaires reported: a few mobile homes fell off their foundations; interior walls sustained a few large cracks; a few windows cracked; many small objects overturned and fell; heavy appliances were displaced by inches; paved sidewalks and streets sustained large cracks and large displacements; an old highway bridge sustained structural damage; ground slumps appeared on hillsides.
	RB	**Rockaway Beach OR.** Postal questionnaires reported: a few windows cracked; a few small objects overturned and fell; heavy appliances were displaced by inches; a few items were shaken off store shelves.
	RE	**Renton.** Buildings on the Boeing campus sustained water damage and damage to ceilings and windows (Nisqually Earthquake Clearinghouse Group, 2001). Several Boeing buildings were red-tagged. Nonstructural damage included filing cabinets that fell over, computers and monitors that were shaken from desks, a fallen ceiling light and some broken water pipes. An internet respondent reported, "People were diving under desks, the lights went out, plaster and tiles were falling from the walls and ceiling and windows were breaking. I decided to head down the stairs. About a third way down another large wave hit and people fell and were thrown aside. I could see big cracks going up the walls of the stairwell and what looked like the top of the stairs starting to separate from the second floor. A quick visual of the outside of the building showed many broken windows." A landslide dammed the Cedar River; engineers breached the landslide to avoid flooding nearby homes (NECG, 2001).
	RY	**Raymond.** Postal questionnaires reported: one entire block of buildings was condemned; exterior walls sustained large cracks and partially collapsed; clay tile sustained large cracks; interior walls sustained a few large cracks and split at seams; a few old chimneys twisted or leaned; a few windows cracked; many small objects overturned and fell. Percentages of effects noted by three internet respondents: difficulty standing or walking during the earthquake— 67%; objects toppled over or fell off shelves—67%; pictures on walls moved or were knocked askew—100%. masonry fell from block or brick walls
	SA	**Saint Helens OR.** Postal questionnaires reported: a few old chimneys cracked, twisted, leaned, lost bricks or fell; a few windows cracked.

DATA TABLE – Earthquake Descriptions

Mercalli Intensity (I-XII)	City Abbr.	Description
	SE	**Seattle.** Many buildings were damaged in Sodo along 1st Ave. South from South Jackson St. to South Hanford St. Along this stretch cracked or collapsed walls, fallen parapets and broken windows were common in old buildings (fig. 12; fig. 13). Notable damage included collapsed unreinforced masonry walls at the Seattle Chocolates building and at the Acme Tool and Specialty building and a collapsed front facade at the Starbucks Headquarters building. Soil liquefaction, as evidenced by ejection of sand, was observed at many locations of the south of downtown Seattle area water pouring out from the sidewalk." The Flentrop Organ in Saint Marks Episcopal Cathedral was heavily damaged (Seattle Post Intelligencer, March 3, 2001, p. A6, col. 1).
	SH	**Shelton.** The USGS field team found sporadic chimney damage throughout the city, with some blocks at the north end of town having many damaged chimneys. Parapets fell on both sides of the old unreinforced masonry Parkview Manor. Objects fell from grocery shelves (several dozen in one aisle). Percentages of effects noted by 30 internet respondents: difficulty standing or walking during the earthquake— 77%; objects toppled over or fell off shelves—77%; pictures on walls moved or were knocked askew—80%; furniture or appliances slid, toppled over, or became displaced —33%; hairline cracks in walls—30%; a few large cracks in walls—17%
	SI	**Sidney B.C.** Percentages of effects noted by three internet respondents: difficulty standing or walking during the earthquake— 33%; objects toppled over or fell off shelves—33%; pictures on walls moved or were knocked askew—33%; hairline cracks in walls—33%; ceiling tiles or lighting fixtures fell—33%. Two of the three noted feeling the quake only.
	SN	**Snoqualmie.** The Snoqualmie–Fall City road was closed due to downslope movement of part of the roadbed (press reports). Percentages of effects noted by 23 internet respondents: difficulty standing or walking during the earthquake— 57%; objects toppled over or fell off shelves—74%; pictures on walls moved or were knocked askew—78%; furniture or appliances slid, toppled over, or became displaced —48%; hairline cracks in walls—52%; a few large cracks in walls—57%; one or several windows cracked—13%; ceiling tiles or lighting fixtures fell—57%; cracks in chimney—13%; major damage to old chimney—9%; masonry fell from block or brick became displaced—57%; hairline cracks in walls—14%; a few large cracks in walls—21%.
	SP	**Spokane.** Felt by most people. Small vibration. No damage.

DATA TABLE – Earthquake Descriptions

Mercalli Intensity (I-XII)	City Abbr.	Description
	TA	**Tacoma.** Postal questionnaires reported: a few old chimneys twisted, leaned, lost bricks or fell; a few windows cracked; small appliances overturned and fell to the floor; a few tombstones twisted or fell. Half a dozen bricks and three coping stones fell from the rear facade of a century-old high school (press reports). A landslide at Salmon Beach damaged nine homes and forced evacuation of others (Nisqually Earthquake Clearinghouse Group, 2001). The Port of Tacoma reported buckled pavement and structural damage to three buildings (Nisqually Earthquake Clearinghouse Group, 2001, March 2001). Merchants in the Tacoma Mall reported "minor damage from toppled shelves, broken glassware and oozing bottles." Fallen items cluttered each of Rite- Aid's aisles (Tacoma News Tribune, March 1, 3001).
	TO	**Tokeland.** Postal questionnaires reported: a few old chimneys cracked; a few windows cracked; a few small objects overturned and fell; several knickknacks broke; many items were shaken off store shelves.
	TP	**Toppenish.** Barely Felt
	WI	**Winlock.** Percentages of effects noted by seven internet respondents: difficulty standing or walking during the earthquake—86%; objects toppled over or fell off shelves—86%; pictures on walls moved or were knocked askew—86%; furniture or appliances slid, toppled over, or became displaced—43%; hairline cracks in walls—43%; a few large cracks in walls—29%.
	WW	**Walla Walla.** Not noticed by most people. Some swinging of objects.

2.6 Shaking during the 2001 Nisqually Earthquake

(Colored Pencils)

15 points

2.6 Shaking during the 2001 Nisqually Earthquake

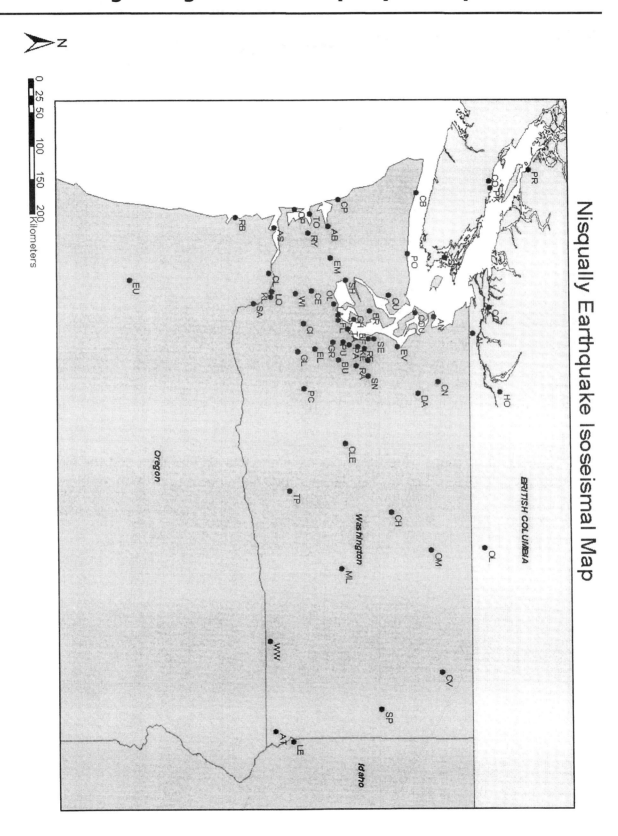

2.6 Shaking during the 2001 Nisqually Earthquake
(Colored Pencils)   15 points

Questions

1. After you have finished making an isoseismal map, please contour Map 2 (on the next page) which shows the measured accelerations from this earthquake. In pencil, please contour the following values: 1, 4, 9, 18, 34, 65, 124.

2. Color between the contours with the following color code:

   Below: 1   Dark Blue
   1 - 4   Light Blue
   4 - 9   Green
   9 -18   Yellow
   18-34   Orange
   34-65   Red
   65 and above Dark red (maroon/brick red)

3. Compare the two maps. Are there places where the acceleration and the Mercalli intensity don't match (i.e. low acceleration but high intensity or high acceleration but low intensity)? Describe where this occurs.

4. What could cause this/these incongruities?

2.6 Shaking during the 2001 Nisqually Earthquake
(Colored Pencils) 15 points

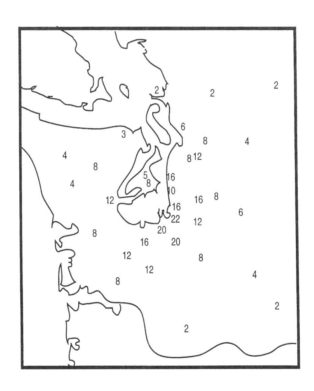

Map 2. Accelerations measured during the Nisqually Earthquake

2.7 Earthquakes in the Pacific Northwest
(Internet Access, Word Processor)   10 points

Name_____

Due Date _____

In this assignment, you will research the sources and types of earthquakes in the Pacific Northwest by visiting two websites:

http://www.geophys.washington.edu/SEIS/PNSN/, then click on "Hazard Summary (PDF)" – it might help to print this two-page PDF document.

http://seattlepi.nwsource.com/local/59909_quake27.shtml, then click on the graphic "Quake scenarios" that occurs within the article. It might be helpful to print this one-page PDF graphic for reference, as well as the article it came from.

NOTE: The information you obtain from this assignment <u>must</u> be incorporated into your Earthquake Hazard and Risk Report in which you describe the hazards of earthquakes within the Pacific Northwest.

Your answers should be <u>typed</u> on a separate sheet of paper.

1.   Describe the general tectonic setting of the Pacific Northwest.

2.   Within our plate tectonic setting, there are three distinct sources of earthquakes. Identify these three earthquake sources and <u>explain</u> where within the plate tectonic setting the ruptures occur and how deep they are.

3.   In what year was the last known "Subduction Zone" earthquake?

4.   What is the recurrence interval for Subduction Zone earthquakes? (i.e. How often, on average, do they occur?)

5.   Which of the three sources describes earthquakes occurring along the Seattle Fault?

6.   When was the last major earthquake on the Seattle Fault?

7.   What is the recurrence interval for shallow crustal fault earthquakes? (i.e. How often, on average, do they occur?)

8.   Which of the three sources/types produce the most frequent earthquakes? How often do they happen? When was the last one?

9.   Which of the three earthquake types can produce the highest magnitude earthquakes?

2.7 Earthquakes in the Pacific Northwest

(Internet Access, Word Processor)   10 points

Examine the "Quake Scenarios: Type determines extent of damage" graphic at the Seattle PI website.

10. Which quake type produces the longest duration of shaking?

11. Which of the quakes will produce the strongest intensity of shaking in the Puget Sound area?

12. In terms of how damage varies by building type, which of the quake types do you think will be the most damaging or catastrophic for the Seattle/Tacoma areas? Support your answer with explanations or reasons for your decision.

13. Return to the main page of the both web sites. Write complete bibliographic references for the websites using the guidelines here:

    Sponsoring Organization, Year of Publication, *Title of Web Site*. [Complete URL], Date Accessed.

2.8 Ground Shaking and Damage at Your House

(Internet/Acrobat Reader, Word Processor)   15 points

Name _____

Due Date _____

In this activity, you will access the ground shaking hazard and potential damage to your home from an earthquake of a certain magnitude.

> **NOTE: The information you obtain from this assignment <u>must</u> be incorporated into your Earthquake Hazard and Risk Report in which you describe your personal risk from ground shaking.**

Finding Peak Ground Acceleration at Your Location

Earthquakes cause the ground surface to move in many different directions. The change of velocity in the ground surface during shaking is called **acceleration**. Buildings (and people) experience acceleration as a force pushing on them. High accelerations are more damaging to buildings then low accelerations and high accelerations lead to higher intensity of damage at a given location. As a result, engineers are very interested in determining the **peak ground acceleration** (PGA) probable in a location, so they can design structures to withstand the potential shaking. Acceleration values are reported in %g of gravity.

1. To find the Peak Ground Acceleration, please visit the following website:

> USGS, 1996, *Probabilitistic Hazard Lookup by Zip Code*.
> Online: http://eqint.cr.usgs.gov/eq/html/zipcode.html

2. Enter the zip code for your home address and press submit. You will receive a report similar to the example below. The value you are interested in is the first number in the left-hand column.

```
The input zip-code is 98198.
    ZIP CODE                        98198
    LOCATION                        47.4003 Lat. -122.3091 Long.
    DISTANCE TO NEAREST GRID POINT  0.6837 kms
    NEAREST GRID POINT              47.4 Lat. -122.3 Long.
    Probabilistic ground motion values, in %g, at the Nearest Grid
point are:
                10%PE in 50 yr   5%PE in 50 yr   2%PE in 50 yr
        PGA        31.293449        40.851151       55.394939
    0.2 sec SA     66.365044       104.852898      127.370201
    0.3 sec SA     59.725670        91.768822      122.026100
    1.0 sec SA     20.979300        29.720091       43.530849
```

73

2.8 Ground Shaking and Damage at Your House

(Internet/Acrobat Reader, Word Processor)   15 points

3. What is your zip code? _____

   What is the PGA value under 10% PE in 50 year? _____ % g

   What does this mean? It means that there is a 10% probability that the Peak
   Ground Acceleration will exceed this value in the next 50 years!

Comparing PGA to the Mercalli Intensity Scale

Let's find out what these acceleration values mean to your home in terms of the
intensity of shaking and potential damage. Start by finding the Mercalli Scale Intensity
Equivalent for the PGA determined above. You can look at the Mercalli Scale at this
website:

   Bolt, Bruce, 1993, Abridged Modified Mercalli Intensity Scale. Online at:
   http://www.eas.slu.edu/Earthquake_Center/mercalli.html
   (Note: There is an underscore in this URL: Earthquake_Center).

Notice that the acceleration values are given in g and not in %g. You will have to move
the decimal point of the acceleration value above two spaces to the right. Example:
31% g = 0.31 g.

4. What Modified Mercalli Scale Intensity does the PGA determined in #3 equate to:

   _____

5. Write the Mercalli Scale description of this shaking.

2.8 Ground Shaking and Damage at Your House
(Internet/Acrobat Reader, Word Processor)   15 points

Potential Damage to Your Home

Maybe you can find out more about the possible damage to your home by looking for information specific to the type of building you live in. The following website provides information on housing types and potential structure damage based on type.

ABAG (Association of Bay Area Governments), 2003, Impacts of California Earthquakes on Buildings from Shaken Awake. Available Online: [http://www.abag.org/bayarea/eqmaps/shelpop/bldg.html]

> Read the classification
> Click on the type of construction that best describes your home.
> Read the description provided.

6. Building type of your home: _____

7. Based on the construction of your home, what structural damage is likely to occur to your home as a result of severe earthquake shaking? Does your home contain any of the weaknesses mentioned in the description? If so, which?

8. Examine the graph: <u>Percent Uninhabitable By Intensity Level</u>. Estimate the percentage of homes (of your building type) that are likely to be uninhabitable after an earthquake of the intensity you found in question 4:

_____

Synthesizing Ground Shaking Information

9. In your Earthquake Hazard and Risk Report, you will be asked to discuss the hazard and risk to you of ground shaking at your home. You now have all the scientific data that you need to make this evaluation. Using the information above, <u>type</u> a paragraph or two that synthesizes the information for your report. You will need to use a separate piece of paper. Be sure to include <u>citations</u> to the sources of your information.

2.9 Earthquake Hazards at Your House
(Internet Access, Word Processor, Camera)   15 points

Name _____

Due Date _____

Earthquakes can generate hazards such as ground shaking (and subsequent damage to property), ground rupture from faulting, liquefaction of soils and sediments, landslides, fire, and tsunami. Which hazards affect you depends on many factors, like the type of construction of your residence, what kind of geologic materials it sits on, if you are near steep slopes, power lines, natural gas pipelines, or large bodies of water like the Puget Sound.

In this activity, you will begin to access some of the hazards associated with earthquake activity and how they will affect you personally. You will be asked to take a look around and photograph your residence, concentrating on areas that might present a problem, and locate yourself and important geological and neighborhood features on a hazard map.

NOTE: The information in this activity includes items that you <u>must</u> discuss and use in your Earthquake Hazard and Risk Paper. It is recommended that you type your answers on a separate paper, as this will serve as the starting point of the Risk Assessment part of your Earthquake Paper.

Characterizing Where You Live and Potential Hazards At Your Residence

1. Describe the location of your home/apartment.
    Example: "My home is located in Ravensdale; a small town about 35 miles southeast of Seattle in King County, Washington."

2. Describe the type of residence you live in. (Attach a <u>photo</u> of the outside of your residence.)
    Example: "I live on the basement floor of a 3-story brick apartment building. The apartments were built in the mid-1970's."

3. Describe the surrounding topographic features. In other words, talk briefly about what the land surface is like in the area surrounding you. (Attach <u>photos</u> of your yard and surrounding area to help you illustrate.)
    Example: "The apartment complex is built on flat ground near Rock Creek, a small stream that flows a few feet from the building. In back of apartment, about 40 feet from the parking lot is a steep, almost vertical bluff. At the top of the bluff, construction of a new housing development is underway,"

4. What type of soils or sediments does your residence sit on? (Remember, you looked this up on an earlier homework! Reiterate what you found here and make a citation to the map where you got the information.)

2.9 Earthquake Hazards at Your House
(Internet Access, Word Processor, Camera)   15 points

5. Chimneys are made of masonry (brick, cinderblock, cement) to prevent houses from burning down when people use their fireplace. Moderate or severe earthquake shaking often causes damage to chimneys because masonry structures are inflexible and do not "give" during shaking. It is often advisable after an earthquake to have your chimney inspected before lighting a fire, even if you cannot see visible damage.

    Do you have a fireplace in your home/apartment? ___Yes   ___No
    (If yes, attach a <u>photo</u> of the outside chimney.)

6. Homes with natural gas may be at risk for fire after an earthquake. It is important to know if you have natural gas coming into your house.

    Do you have natural gas in your home/apartment? ___ Yes   ___No
    (If yes, locate and <u>photograph</u> the shut-off valve and describe how to turn off the gas.)

7. You should also know how to shut off electrical power to your home. Find and <u>photograph</u> your electric panel and describe how you would shut-off power to your home.

8. Water heaters are another problem area during earthquake shaking. They are heavy and connected by inflexible piping and can detach from the wall.   As a result, it is recommended that water heaters be strapped down. Locate and <u>photograph</u> your water heater and describe how to turn it off.

    Is your water heater strapped down?   ___Yes   ___No
    Is your water heater gas or electric?   ___ Gas   ___Electric

9. Moderate to strong earthquake shaking can cause even heavy objects to fall and cause damage and/or injury. Look around your bedroom. Attach a couple of <u>photographs</u> and describe anything that might present a hazard to you personally if you were in your bedroom during an earthquake.

10. Do you have an earthquake kit?   ___Yes   ___No

    If so, where is it located within your home/apartment?

2.9 Earthquake Hazards at Your House

(Internet Access, Word Processor, Camera)   15 points

Evaluating Other Earthquake Hazards

To answer the following questions, you will be using the following resource:

USGS, 1999, *Lifelines and Earthquake hazards in the greater Seattle area*, Open-File Report 99-387.

This map is available in the classroom, in the school library, and on-line at: http://geology.wr.usgs.gov/wgmt/feature/lifeline/aboutmap.html

11. **Ground Rupture**. People living on or very near a fault can experience damage due to the ground rupturing and moving along the fault.

    Do you live within the zone of possible ground rupture along the Seattle Fault?

    ___ Yes   ___No

    If no, how many miles is the rupture zone from your residence? _____ miles. (Use the scale on the map to find the distance.)

    Do you live within the zone of possible ground rupture along the Tacoma Fault? **(NOTE: You will have to use the map in classroom to answer this question – it is the only one with the fault drawn on it.)**

    ___ Yes   ___No

    If no, how many miles is the rupture zone from your residence? _____ miles. (Use the scale on the map to find the distance.)

12. **Liquefaction**. Liquefaction is the conversion of apparently stable, cohesive sediment into a liquid slurry. Where soils/sediment are prone to liquefaction, buildings, structures, and roadways can sink in, tilt, or collapse. The map shows areas of unconsolidated young geologic deposits that have moderate to high liquefaction potential.

    Do you live in the area susceptible to liquefaction?   ___Yes   ___No

    Note: This is a simplified map. There are other more detailed liquefaction maps available for many areas in the school library (or in class). When you write your Earthquake Hazard and Risk Report, you <u>will</u> want to consult a more detailed map, if available for your area.

2.9 Earthquake Hazards at Your House
(Internet Access, Word Processor, Camera)   15 points

13. **Fire**. Earthquakes can damage natural gas and fuel oil pipelines, as well has high voltage electrical lines, which can lead to fire and/or explosions. Use the scale on the map to determine the distance from your residence to the closest:

        Liquid fuel pipeline   _____ miles
        Natural gas pipeline   _____ miles
        500 kV electric line   _____ miles

14. **Tsunami**. An large earthquake along the Seattle Fault could generate tsunami waves in the Puget Sound and cause damage to property or injury/death to people who live near the water at low (< 20 feet) elevations.

    Is your home near the Puget Sound at a low elevation? ___Yes   ___No

    If no, how far is your house from the Puget Sound? _____ miles

15. **Landslides.** Earthquake shaking can loosen material on steep or unstable slopes and produce landslides, which can lead to damage or property or even injury/death to people.

    Do you live at the top of a steep hill?   ___Yes   ___No
    Do you live at the base of a steep hill?   ___Yes   ___No
    Do you live in the middle of a steep hill? ___Yes   ___No
    What is the approximate distance of the closest hill?   _____ miles

16. Type a complete bibliographic reference for the Lifeline hazards map. When writing your paper, what is the proper way to make a citation for this map?

2.10 Forecasting Future Earthquakes

(Internet Access, Ruler)   20 points

(Eric Baer, Highline Community College, 2003)

Your Name _____

Group Member _____

Group Member _____

Group Member _____

Conditional probabilities are estimates of the probability of an earthquake of given magnitude occurring in an area within a specified time period. These estimates are based on a synthesis of real data, historical records and geologic evidence of pre-historic earthquakes.

For this activity, you will use the history of earthquakes in the Pacific Northwest (PNW) in an attempt to determine the hazards we might be facing in the future.

You should work in groups of 5 people to divide up the work and share the information. You will first be given a homework assignment to gather data and then come to class to analyze that data.

In-class, you should use a calculator. The math is very simple multiplication, division, and finding averages; however, the numbers are not even and you should carry at least 3-4 numbers out from the decimal. You should also show your math work. If you don't I can't see where or how you went wrong (if you do) and can't give you partial credit.

Follow the steps in order. This will lay the larger problem out for you in a meaningful way.

Reminder: Even though you are working with a group to you should each be filling in your data sheet, making your own graph, and answer the discussion questions in your own words.

2.10 Forecasting Future Earthquakes

(Internet Access, Ruler)   20 points

Section 1. Homework (10 points)

You will be using the USGS to help you find the number of earthquakes in a given number of years within a limited magnitude range.

1. At home, go to http://neic.usgs.gov/neis/epic/epic_rect.html
   Note: There is an underscore in the URL: epic_rect
2. Select " 3. Screen File Format (80 columns)"
3. Check "USGS/NEIC (PDE) 1973 – Present"
4. Input the following data to limit your search to the Pacific Northwest:

50	Top latitude of rectangle
42	Bottom Latitude of Rectangle
-119	Right Longitude of Rectangle
-127	Left Longitude of Rectangle

5. Input the appropriate starting date and ending date for the data you want and the appropriate magnitude range. Leave the other fields blank.

6. In the next three tables, each member of your group will be responsible for filling out one line of data. For instance, John might be responsible for 1988-1990 M3 earthquakes and 1979-1983 M4 quakes while Amy is responsible for the second line in each table. No one person will be all the data in these tables. **YOU MUST COME TO CLASS WITH THIS DATA OR YOUR GROUP WILL NOT BE ABLE TO FINISH THE ASSIGNMENT.**

Three-year period	Number of M. 3.0-3.9 Earthquakes
1990-1992	
1993-1995	
1996-1998	
1999-2001	
2002-2004	
Total in 15-year period	

Five-year period	Number of M. 4.0-4.9 Earthquakes
1980-1984	
1985-1989	
1990-1994	
1995-1999	
2000-2004	
Total in 25-year period	

2.10 Forecasting Future Earthquakes

(Internet Access, Ruler)   20 points

All students should complete the following:

7. How many M5.0-M5.9 Earthquakes were there between 1973 and 2004?

32-year period	Number of M. 5.0-5.9 Earthquakes
Total in 32-year period	

Finally, go to http://www.pnsn.org/HIST_CAT/catalog.html and answer the following questions from the table there. (Note: This URL has an underscore: HIST_CAT)
Be careful! The list has magnitudes outside of the range you are interested in!

8. How many M6.0-6.9 earthquakes have there been in the Pacific Northwest since 1872?

9. How many M7.0-7.9 earthquakes have there been in the Pacific Northwest since 1872?

2.10 Forecasting Future Earthquakes

(Internet Access, Ruler)

20 points

In Class Activity (10 points)

Gather with your group and share your data. Fill out the tables and answer the following questions.

Three-year period	Number of M. 3.0-3.9 Earthquakes
1990-1992	
1993-1995	
1996-1998	
1999-2001	
2002-2004	
Total in 15-year period	

1. What is the average number of earthquakes per year between magnitude 3.0 and 3.9? (Hint: Average = Total/15) Record your answer in the attached data table sheet.

2. What is the number of M 3.0-3.9 earthquakes expected in ten (10) years? (Hint: # of quakes expected in 10 years = Average # of quakes per year x 10 years). Record your answer in the attached data table sheet.

3.   What is the average time (in days) between magnitude 3.0-3.9 earthquakes? (Hint: Average time between quakes = 365 days per year/ Average # of quakes per year). Record your answer in the attached data table sheet.

4. What are some problems that you might see with the above data? How significant might these errors be? *Be specific and detailed in your answer. Use another piece of paper or the back.*

2.10 Forecasting Future Earthquakes
(Internet Access, Ruler)   20 points

5. Fill out the table and calculate:

Five-year period	Number of M. 4.0-4.9 Earthquakes
1980-1984	
1985-1989	
1990-1994	
1995-1999	
2000-2004	
Total in 25-year period	

a) How many M. 4.0-4.9 earthquakes occur each year on average? How many would you expect in 10 years?

b) What is the average time period (in days) between quakes?

c) Look at the data in the above table again. Between March 24 and May 19, 1980 there were over 200 earthquakes between M. 4.0-4.9. Why do you think this is?

d) Should you include 1980 data in your estimates or disregard it? **Why or why not?** *Please fully discuss the reasons for your answer, being sure that your reasons either support including or disregarding this data.*

2.10 Forecasting Future Earthquakes
(Internet Access, Ruler)   20 points

6.   We will move on to M5.0-5.9 Earthquakes now.

a)   How many were there during the period you looked at? (Use your answer above to determine what you should do about 1980.)

b)   How many years were there?

c)   On average, how many M. 5.0-5.9 occur in a year?

d)   How many would you expect in 10 years?

e)   What is the average time period (in days) between quakes?

7.   Use the rest of your data to completely fill in the following results table

DATA TABLE SUMMARIZING RESULTS FROM 1 - 5

Magnitude Range	Average number in one year (b)	Number Expected in ten years (c)	Time between quakes (days) (d)	Time between quakes (years)
3.0 – 3.9				
4.0 – 4.9				
5.0 – 5.9				
6.0 – 6.9				
7.0 – 7.9				

8.   Plot the magnitude against the average amount of time (in years) between each quake. A piece of graph paper is provided. You can plot all magnitude 3.0-3.9 earthquakes as a point on the vertical line labeled 3 etc.

2.10 Forecasting Future Earthquakes
(Internet Access, Ruler)   20 points

9. From your data, project how often magnitude 8 and 9 earthquakes might happen. (**You do this by extending the best-fit line**.) What is their average rate of reoccurrence?

10. Recent evidence has shown that there was a great (M 8-9) earthquake 300 years ago and 1100 years ago. Does this data fit with your calculations?

11. For every great earthquake (M 8-9) that occurs in this area, how many M 6-6.9 earthquakes would you expect?

12. Given the above information, your group should discuss the following question: Should public officials, emergency planners, building engineers, and/or individuals focus earthquake prevention and preparedness efforts on M. 8-9 earthquakes or M. 6-7 earthquakes? Explain your reasoning in detail. What are the positive and negative implications of your choice?

2.10 Forecasting Future Earthquakes
(Internet Access, Ruler)   20 points

Magnitude vs. Period

Period (Years Per Earthquake)

Magnitude

2.11 The Seattle Fault Scenario

(Library, Word Processor)   20 points

Name _____

Due Date _____

NOTE: The information you obtain from this assignment should be incorporated into your Earthquake Hazard and Risk Report as part of the regional hazard and risk assessment. It is information about the economic, physical, and human impacts of a large earthquake in the Puget Sound area.

Introduction

In 2005 a team of engineers, geoscientists, emergency and risk managers, hazard mitigation specialists, and business people completed an "earthquake scenario" for the Seattle Fault. The purpose of this 3-year project was to prepare a credible description of the damage and impacts likely to result from a large earthquake along the Seattle Fault. Although this scenario vividly describes impacts and damages to the Seattle area, descriptions are for a HYPOTHETICAL earthquake. The scenario document will assist citizens, elected officials, business owners, emergency managers, land use planners, engineers and others to prepare for an appropriate response when such an earthquake actually occurs. The results of their efforts can be seen in:

> Stewart, M. (ed.), 2005, *Scenario for a Magnitude 6.7 Earthquake on the Seattle Fault,* EERI and EMD (publishers), pp. 162.

Several copies of this publication are on reserve at the Circulation Desk in the school library.

Your assignment is to read the Executive Summary of the Scenario (p. 1-15) and answer the following questions. Although you are not specifically asked to read the entire publication, it will be helpful for you to skim the remainder of the document. Some of the following questions refer specifically to graphics/tables in the document. Your responses should be typed and submitted on a separate paper.

The Scenario Earthquake

1. The scenario describes the potential impacts that may result from the occurrence of large earthquake along the Seattle Fault. Which of the three types/sources of Pacific Northwest earthquakes is described in this scenario? (You might want to refer back to activity 2.7 to refresh your memory on the three types.)

2. There are other faults of this type in the Puget Sound area. Name 2 other faults in the Puget Sound that could potentially produce impacts similar to those in this scenario.

2.11 The Seattle Fault Scenario

(Library, Word Processor)   20 points

3.   What is the estimated probability of occurrence in 50 years for a Seattle Fault M 6.5 or greater earthquake? What is the approximate recurrence interval? (Table I-1, p. 18.)

4.   If an earthquake doesn't occur on the Seattle Fault for 999 year, what is the probability of occurrence the next year? (Think back to the lecture on hazard and risk.)

5.   The scenario earthquake is smaller than a real earthquake that occurred along the Seattle fault about 1,100 years ago (in terms of its magnitude, length of rupture and displacement). What is the magnitude, length of rupture and maximum displacement of the ground surface modeled in this scenario?

6.   How does the scenario event (Seattle Fault M 6.7) compare with what the area experienced in the Nisqually, 2001 M 6.8 earthquake? (Table E-1, p. 4)

7.   Describe two (2) significant ways in which utilities (electricity, water, wastewater and treatment, natural gas and liquid fuels, and communications) will be impacted by the scenario earthquake.

8.   Describe two (2) significant ways that the scenario earthquake will impact our regional transportation system.

9.   Describe two (2) significant ways that the scenario earthquake will impact buildings in the region.

10.   Describe three (3) significant economic impacts that are likely to result from the scenario earthquake.

11.   Identify three (3) environmental problems that could result from the scenario earthquake.

2.11 The Seattle Fault Scenario
(Library, Word Processor)   20 points

12. Examine Figure 1-8 (p. 34): NEHRP soil maps for the study region. Describe which areas have soils most prone to liquefaction and state if you live on these soils.

13. Examine Figure 1.9 (p. 35): Peak Ground Acceleration for Seattle Fault Scenario Earthquake using soil maps. Where do you live? (city) _____. Estimate the location of your home on the map. What is the approximate Peak Ground Accelerations your house is likely to experience during this M 6.7 event?

14. For the Peak Ground Acceleration determined above, what is the Modified Mercalli Intensity equivalent (See Table 1.1, p. 25.)? What amount of shaking is likely to occur? Describe some of the potential damage,

15. What is meant by the term "Lifelines" (Chapter 3, p. 51)? Why are lifelines considered important?

16. Examine the maps in Figures 3.1 (all lifelines, p. 52), 3.2 (water pipelines, p. 54), 3.4 (sewer lines, p. 56), and 3.7 (natural gas and liquid fuel pipelines, p. 61). All maps show one or more of the region's lifelines plotted on a map with Peak Ground Acceleration data and the modeled fault rupture. Identify and describe the relationship between the location of lifelines to the fault rupture and liquefaction prone areas.

17. After reading the scenario document, do you personally feel you or your community is prepared and can respond to the damage created by a M 6.7 earthquake along the Seattle Fault, or one of the areas other shallow crustal faults? Cite specific examples to support your point.

18. Type a correctly formatted bibliography for the Scenario document to include with your list of references for your paper!

2.12 Finding Patterns in Earthquakes
(Data and maps supplied in class)   10 points

Instructions

Divide into groups of two. Each group will get one of the following sets of data assigned by your instructor:

> Data Group A: Large and very large earthquakes
> Data Group B: Deep Earthquakes
> Data Group C: Recent earthquakes

Each group will be plotting their data on a copy of the map from the following location: http://www.eduplace.com/ss/maps/pdf/world_country.pdf. Please mark these earthquakes off as you plot them. When completed, please pass the list and map to another group.

Data Group A: There will be two groups assigned to Data Group A. Both groups should plot 6 very large and 10 large earthquakes on the map. Please use a red pen for plotting the very large quakes and an orange pen for plotting the large quakes.

Data Group B: There will be two groups assigned to Data Group B. Both groups should plot 25 deep earthquakes on the map. Please use a blue pen for plotting the deep quakes.

Data Group C: All other groups will plot recent earthquakes. Plot 25 earthquakes using a green pen on the maps and pass the map on to another group.

Repeat with another map so by the end of the class period we will have all of the data plotted.

When the plotting is done, we will examine patterns in earthquake distribution.

2.12 Finding Patterns in Earthquakes

(Data and maps supplied in class)   10 points

Questions

Please answer the following questions using SPECIFIC EXAMPLES.

1.   Are earthquakes evenly distributed around the globe?

2.   What areas seem to have unusually high numbers of earthquakes?

3.   What areas are devoid of seismicity?

4.   Look at a globe or other map of the world. Are there features or other characteristics that are also common in earthquake prone regions?

2.12 Finding Patterns in Earthquakes
(Data and maps supplied in class)   10 points

5.   Do all places that have earthquakes have deep earthquakes?

6.   Can you discern a pattern for the deep earthquakes?

7.   Do all places that have earthquakes have large or very large earthquakes?

8.   Can you discern a pattern for the large or very large earthquakes?

9.   Is there a correlation between deep and large earthquakes? Be specific!

2.13 Video: Cascadia - The Hidden Fire

10 points

Name _____

Due Date _____

(Cascadia: The Hidden Fire is produced by Global Net Productions and is available on reserve at the Circulation Desk in the Library.)

The following questions are summary questions. You will need to take notes during the video to answer the questions, so read the questions in advance. Your answers should provide sufficient detail and should be formulated in complete sentences of paragraphs. Please type your responses on a separate paper to turn in to your instructor.

1. The Cascadia region faces a "seismic triple threat". What do geologists mean by this statement? What are the specific sources of this threat?

2. Scientists in the 1980's were unsure if the Cascadia Subduction Zone (CSZ) was active and could produce large "super" earthquakes. Summarize the two conflicting theories about earthquake occurrence in the CSZ and describe the main piece of scientific evidence found to determine which of the theories was correct.

3. Describe the scientific evidence found to support the idea that the last great Cascadia Subduction Zone earthquake occurred about 300 years ago (in 1700 AD).

4. The "Decade of Terror" hypothesis suggested that the Cascadia Subduction Zone breaks in sections and produces several major earthquakes over a relatively short period of time (a decade). The "Instant of Terror" hypothesis suggested that the entire 700 miles of the CSZ breaks and produces one great magnitude 9 earthquake. Outline how scientists determined which of these hypotheses were correct and what the results showed.

5. What are slow silent earthquakes? How often do they occur? How much energy is released by the events? Do they increase or decrease the chances of a large earthquake along the subduction zone?

6. What type of earthquake was the 2001 Nisqually earthquake? How are these earthquakes caused? What are the dates (years) of two similar earthquakes that caused significant damage to the Puget Sound area?

7. What type of earthquake presents the worst case scenario for cities like Seattle, Tacoma, and Portland? How frequently does this type of earthquake occur? How long ago did the Seattle Fault last rupture? Discuss two ways in which Seattle will be impacted economically if one of these events were to occur today.

2.13 Video: Cascadia - The Hidden Fire

10 points

8. Describe a few positive sides to the geologic forces that affect the Cascadia region.

9. You decide to use information from this video in your Earthquake Hazard and Risk paper. Write a correctly formatted bibliographic reference for the DVD-video. Use the Style Guide in this book (6.1) to help you format the bibliography.

3.1 Volcanic Hazard Maps
(Internet Access, Word Processor)   15 points

Name _____

Due Date _____

There are five potentially active or active volcanoes in Washington State. The hazards presented by these volcanoes have been extensively mapped and documented by the USGA. In this assignment, you will be reviewing hazard maps for each of these five volcanoes to determine the risk of these hazards to you at your residence.

The hazard maps are available online at the given websites (complete bibliographies for the websites are listed for you). The maps are also available to you in the school library. If you are using the online map versions, please allow the files time to load. The maps are in PDF format and are very large. Once loaded, you may have to zoom in or enlarge the image to see it better. All maps have a legend that indicates what the colors and patterns mean.

NOTE: The information you obtain in this activity <u>must</u> be included in your Volcano Hazard and Risk Paper, where you are to discuss your personal risk from volcanic hazards.

<u>Mount Adams</u>
Scott, W. E., Iverson, R.M., Vallance, J.W., and Hildreth, W., 1995, *Volcano Hazards in the Mount Adams Region, Washington;* USGS Open-File Report 95-492. Online: http://vulcan.wr.usgs.gov/Volcanoes/Adams/Hazards/OFR95-492/framework.html

1.   Is your residence located within the mapped area?   ___Yes   ___No
     (If so, are you within any of the hazard zones? Which one(s)?)

2.   Do any cities (urbanized areas) fall within the hazard zones? ___ Yes   ___No

3.   If so, identify <u>two</u> of the cities and state the hazard posed to them.

3.1 Volcanic Hazard Maps

(Internet Access, Word Processor)   15 points

Mount Baker

Gardner, C.A., Scott, K.M., Miller, C.D., Myers, B., Hildreth, W. and Pringle, P.T. 1995, *Potential Volcanic Hazards from Future Activity of Mount Baker, Washington:* USGS Open-File Report 95-498. Online:
http://vulcan.wr.usgs.gov/Volcanoes/Baker/Hazards/OFR95-498/framework.html

4.   Is your residence located within the mapped area?   ___Yes   ___No
     (If so, are you within any of the hazard zones? Which ones?)

5.   Do any cities (urbanized areas) fall within the hazard zones? ___ Yes   ___No

6.   If so, identify two of the cities and state the hazard posed to them.

Glacier Peak

Waitt, R.B., Mastin, L.G., and Begét, J., 1995, *Volcanic-Hazard Zonation for Glacier Peak Volcano, Washington:* USGS Open-File Report 95-499. Online:
http://vulcan.wr.usgs.gov/Volcanoes/GlacierPeak/Hazards/OFR95-499/framework.html

7.   Is your residence located within the mapped area?   ___Yes   ___No
     (If so, are you within any of the hazard zones? Which ones?)

8.   Do any cities (urbanized areas) fall within the hazard zones? ___ Yes   ___No

9.   If so, identify two of the cities and state the hazard posed to them.

3.1 Volcanic Hazard Maps
(Internet Access, Word Processor)   15 points

Mount St. Helens

Wolfe, E.W. and Pierson, T.C., 1995, *Volcanic-Hazard Zonation for Mount St. Helens, Washington, 1995:* USGS Open-File Report 95-497. Available http://vulcan.wr.usgs.gov/Volcanoes/MSH/Hazards/OFR95-497/framework.html

10. Is your residence located within the mapped area?   ___Yes   ___No
     (If so, are you within any of the hazard zones? Which ones?)

11. Do any cities (urbanized areas) fall within the hazard zones? ___ Yes   ___No

12. If so, identify two of the cities and state the hazard posed to them.

Mount Rainier

Hoblitt, R.P., J.S. Walder, J.S., Driedger, C.L., Scott, K.M., Pringle, P.T., and Vallance, J.W., 1998, *Volcano Hazards from Mount Rainier, Washington*; USGS Open-File Report 98-428. Online: http://vulcan.wr.usgs.gov/Volcanoes/Rainier/Hazards/OFR98-428/framework.html

Notice the hazard map is divided into two plates. Plate 1 is the main map hazard map. The northern part of the map is continued on Plate 2.

13. Do you live in a lahar inundation zone?   ___Yes   ___No

14. If so, which case?   ___Case 1   ___Case 2   ___Case 3   ___Case M

     Write the description of the case lahar that you live in here.

15. If not, how close is the nearest lahar inundation zone from your house? (The map has a scale!)

                         _____ miles.

3.1 Volcanic Hazard Maps

(Internet Access, Word Processor)   15 points

16. What major life activities (work, school, shopping, etc.) of you, your family and friends take place within the lahar inundation zone?

Lastly, look at the following maps on Plate 2.

17. Do you live in the pyroclastic flow hazard zone?   ___Yes   ___No

    If no, approximate the distance from your residence to this zone: _____ miles.

18. Do you live in the lateral blast hazard zone?   ___Yes   ___No

    If no, approximate the distance from your residence to this zone: _____ miles

On a separate sheet of paper, <u>type</u> responses to the following questions.

19. Briefly discuss the direct and indirect effects of lahars to you personally based on where you live, what activities you participate in, etc. (This discussion can be used in your Volcanic Hazard and Risk Paper.)

20. Submit typed bibliographies for all of the maps used in this assignment. You will need to add these to your works cited page of your report when you discuss the hazards of Washington's five volcanoes!

104

3.2 Volcano Movie Review
(Video, Word Processor)   15 points
(Adapted from Gilbert, L., 2002)

Due Date: _____

Introduction
Volcanic eruptions have been often portrayed in the movies, sometimes very accurately and other times quite poorly. Your assignment is to find a movie centered around a volcano and write a maximum of 2-pages (typed, double spaced, 12 point font) reviewing the geologic content of the movie. If you are unsure about the appropriateness of your movie choice, check with me first. Please follow the guidelines below.

Content
1. Introduce the movie you chose.
    State the title and release date. Summarize what geologic processes were central to the movie. Do not re-tell the plot.
2. Volcanic Landforms and Plate Tectonics
    Give the general location of the movie (e.g. Los Angeles, Cascades, etc.) and describe the real tectonic setting of the area (you may have to do some research for this). Then, identify and describe the volcanic landform and the eruptive style(s) shown in the movie. Did the producers select the correct setting for the film? How likely is it that a volcano would erupt in this setting? Was the correct volcano type used in the correct geologic setting?
3. Eruptive Styles, Hazards and Risks
    For the type of volcano portrayed, how scientifically accurate were the geologic hazards portrayed in the film? Did the volcano erupt appropriately? Were the risks associated with this volcano adequately assessed? What flaws did you find? How would you correct those flaws to properly portray volcanic eruptions, hazard and risk?
4. Scientists and Public Officials
    How does the movie depict scientists and how they actually do what they are trained to do? How are issues of prediction, warning, public emergency preparedness, and mitigation portrayed? How do scientists and public officials interact in the movie? Do you think this an accurate portrayal? Why or why not?

Style
Like all other papers for this class, you need an introduction and conclusion. Please use complete sentences, spell check, and proofread for grammar. Use formal English. Write in your own words. Cite the movie, where appropriate, and other references you consult.

Grading
You will be graded on (a) how completely and accurately you follow the instructions (an appropriate choice is part of this!), (b) the coherency and completeness of your paper, (c) grammar and flow (how easy it is to read your paper), and (d) the overall thoughtfulness of your paper (give evidence to support your statements!).

Some ideas (but don't limit yourself to these!):
Volcano (1997), Dante's Peak (1997), Volcano, Fire on the Mountain (1997), Joe Versus the Volcano (1990), The Devil at 4 O'clock (1961), Journey to the Center of the Earth (1959). Most of these are available through the King County Library System.

3.3 Disaster in Armero
(Word Processor, Article distributed in class)   10 points

Name _____

Due Date _____

This article, "Eruption in Columbia" appeared in the May 1986 issue of National Geographic. Additional information from class and the video Understanding Volcanic Hazards may be helpful.

On a separate sheet of paper, type a short paragraph on **each** the following:

1. Describe the volcanic events that occurred at Nevado Del Ruiz in 1984.

2. Describe what human errors led to these events to becoming a catastrophe.

3. What are the similarities to the situation at Mount Rainier and Armero before the disaster?

4. What lessons could be learned from this disaster and what should we do to prevent a similar occurrence at Mount Rainier?

5. Type a correctly formatted bibliography for the National Geographic article.

3.4 Ash Fall Hazards at Your Home
(Word Processor)   10 points

Name _____
Due Date _____

In this assignment, you will be evaluating the volcanic ash fall hazard at your home. To do this, we will be using a Volcanic Hazard Map for the Cascades created by the USGS. The map is from the report:

Hoblitt, R.P., J.S. Walder, J.S., Driedger, C.L., Scott, K.M., Pringle, P.T., and Vallance, J.W., 1998, *Volcano Hazards from Mount Rainier, Washington*; USGS Open-File Report 98-428. Online: http://vulcan.wr.usgs.gov/Volcanoes/Rainier/Hazards/OFR98-428/framework.html

Please examine the maps carefully and answer the questions on the following page.

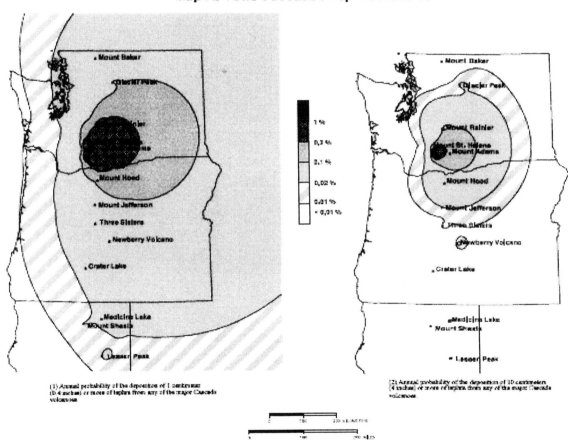

109

3.4 Ash Fall Hazards at Your Home

(Word Processor)

10 points

3.4 Ash Fall Hazards at Your Home

(Word Processor)   10 points

1. Mark the location of your home on the above maps.

2. Do you live in a tephra hazard zone for all of the Cascade volcanoes?

          ___ Yes   ___No

3. What is the probability of 1 cm of ash falling on your house this year? In 60 years?

4. What is the probability of 10 cm of ash falling on your house this year? In 60 years?

Please type the responses to questions 5 and 6 on a separate piece of paper and turn them in. **NOTE: This is valuable information that should be included in Volcanic Hazard and Risk paper!**

5. What would be the effects of receiving 1 cm of ash in the area where you live?

6. What would be the effects of receiving 10 cm of ash where you live?

3.5 Video: Understanding Volcanic Hazards

10 points

Name _____

Due Date _____

Video questions based on video "Understanding Volcanic Hazards" (available on reserve at the Circulation Desk in the school library).

1.   List the 7 main volcanic hazards

2.   What are 3 possible dangerous effects of ash falls (usually called *pyroclastic falls).*

3.   In which ways are hot ash flows (also called pyroclastic flows) deadly? (In other words, what would happen to you if you are caught in the path of a *pyroclastic flow?*)

4.   How fast can pyroclastic flows travel?

5.   What are the differences between pyroclastic flows and volcanic mudflows (also called *lahars*)?

6.   How fast can lahars travel?

3.5 Video: Understanding Volcanic Hazards

10 points

7.   Name a volcano that was decapitated by a massive volcanic landslide (also called *debris avalanche*).

8.   How fast can debris avalanches travel?

9.   What are two ways volcanoes can generate tsunami waves?

10.   What type of impact can lava flows have on humans?

11.   Name 2 dangerous volcanic gases.

12.   Cite an example of a recent tragedy causes by the emission of CO_2 gas.

3.6 Video: Kilauea, Close Up of an Active Volcano

10 points

Name _____

Due Date _____

Video questions from the video: "Kilauea, Close Up of an Active Volcano", available on reserve at the Circulation Desk in the school library. (BAER Media 551.21 K48 1994).

1. Which is the oldest island among the inhabited ones of the Hawaiian archipelagos and how old are the oldest rocks on that Island?

2. How old are the oldest rocks on the Big Island of Hawaii?

3. When did Mauna Loa last erupt?

4. Which is the most active volcano on the Big Island of Hawaii?

5. What are the 2 areas from which Kilauea volcano erupts?

6. When did the last series of eruptions at Kilauea begin?

7. How did they try to stop the lava from destroying the Visitor Center? Were they successful?

8. How many homes were destroyed by the lava flow since the eruption began?

9. How much economical damage (in $$$) did this last series of eruptions cause?

10. What are the properties of Hawaiian lavas?

11. What is the most explosive type of volcanic activity in Hawaii?

3.6 Video: Kilauea, Close Up of an Active Volcano

10 points

12. How is the movement of lava on the surface of a lava lake similar to the movement of tectonic plates?

13. What are the most common gases in Hawaiian lavas?

14. What are the main differences between pahoehoe and aa lava?

15. Can aa lava revert to pahoehoe lava?

16. What happens when a pahoehoe flow finds an obstacle in its path?

17. What happens when aa flow finds an obstacle (such a car) on its path?

18. How do lava tubes form?

19. How high can it be the temperature of lava in a lava tube?

20. How do ropy folds develop in pahoehoe lava?

21. What happens when the lava enters the water slowly?

22. What happens when lava enters the water very fast?

23. Where and how do pillow lavas form?

3.7 Video: Perilous Beauty

10 points

Name _____

Due Date _____

The video "Perilous Beauty" is available on reserve at the Circulation Desk in the school library.

1. Which volcano did Crandell and Mulleneux forecast (in 1975) would be the next volcano to erupt in the Cascades?

2. They gave a date before which it was likely to erupt. What was the date?

3. When did this volcano actually erupt?

4. Were they correct in their prediction?

5. What volcano do volcanologists now view as the most dangerous in the Cascades?

6. How long has Mt. Rainier been without an eruption?

7. What type of deposits did Crandell expect to find around Enumclaw?

8. What did he actually find?

9. What is the name of the large lahar (mudflow) that Crandell studied?

10. Where did it come from?

3.7 Video: Perilous Beauty

10 points

11. What event caused geologists to realize that volcanoes could have very large landslides?

12. What does the landslide spawn that then moves farther downstream?

13. Name one other place (besides Mount St. Helens and Mt. Rainier) that there has been a large landslide giving hummocky topography.

14. How deep was the Osceola lahar?

15. How fast did it move?

16. Where did the Osceola finally end?

17. When did it happen?

18. What happened to the giant crater that was left in Mount Rainier after the Osceola event?

19. How many other immense flows have occurred since the Osceola 5700 years ago?

20. What flow came off of the Sunset amphitheater about 500 years ago?

21. What town sits on the Electron flow?

22. How long would it take for a flow similar to the Electron Lahar to get to Orting?

118

3.7 Video: Perilous Beauty

10 points

23.   Was there a volcanic eruption associated with the Electron event?

24.   What does this mean in terms of warning time for a future event?

25.   What could trigger a massive landslide on Mt Rainier?

26.   How often do large landslides occur on Mt. Rainier?

27.   On the upper slopes of Mt. Rainier, what is between each of the lava flow layers?

28.   What does hydrothermal alteration do to solid rock?

29.   What 2 things are needed for hydrothermal alteration?

30.   Where on the mountain is the largest dike and therefore the most amount of hydrothermal alteration?

31.   What is the most far-reaching hazard at Mount Rainier?

32.   What else can cause a lahar on Mount Rainier?

33.   Each ash layer indicates a pyroclastic flow event. How many pyroclastic flow events have occurred since the Osceola 5700 years ago?

3.7 Video: Perilous Beauty

10 points

34. How can Mount Rainier produce smaller, more frequent floods?

35. Since 1968, how many lahars have here been down Tahoma creek?

36. Do most people have to worry about glacial flood generated lahars?

37. What is the chance of a large landslide on Mount Rainier in your lifetime?

38. How many people died in the 1985 eruption of Nevado Del Ruiz in Columbia?

39. How much earlier had Armero been destroyed by lahar?

40. What can you do? (List all 5)

3.8 Video: Kilauea, Lava Flows

10 points

Name _____

Due Date _____

The video "Lava Flows and Lava Tubes" from Volcano Video Productions is currently only available for viewing in class.

1.   What are the main differences between pahoehoe and aa lava?

2.   How do lava tubes form?

3.   Where and how do pillow lavas form?

4.   Where would you find a lava lake?

5.   How is the movement of lava on the surface of a lava lake similar to the movement of tectonic plates?

3.9 Phases of Eruption of Vesuvius, 79 AD

15 points

(Authored by Emanuela Baer, Shoreline Community College)

Name: _____

Due Date: _____

Introduction

The 79 AD eruption of Vesuvius has become in modern volcanology the prototype of large explosive volcanic eruptions beginning with high vertical columns.

Assignment

1. Read the attached letters of Pliny the Younger to Tacitus describing the circumstances of the death of his uncle, Pliny the Elder during the 79 AD eruption of Mt. Vesuvius. Pliny the Younger was a witness and survivor of this eruption. The eruption was called "Plinian" after Pliny the Elder who was killed in it.

2. While you are reading, refer to the modern interpretation of the phenomena occurred during this eruption (which we discussed in class), and try to recognize from the description of Pliny the various volcanic phenomena occurred in the different phases of the eruption. Mark them on the handouts with a highlighter or underlined them and write a margin note next to the correspondent passage with the current name of the phenomenon.

3. On a separate sheet of paper re-write your own description of the eruption (at least one page long) trying to explain what happened integrating the information from the letters of Pliny and your class notes and interpreting it according to the volcanological knowledge of our time.

3.9 Phases of Eruption of Vesuvius, 79 AD

15 points

(Authored by Emanuela Baer, Shoreline Community College)

1. Pliny Letter 6.16

My dear Tacitus,

You ask me to write you something about the death of my uncle so that the account you transmit to posterity is as reliable as possible. I am grateful to you, for I see that his death will be remembered forever if you treat it [sc. in your Histories]. He perished in a devastation of the loveliest of lands, in a memorable disaster shared by peoples and cities, but this will be a kind of eternal life for him. Although he wrote a great number of enduring works himself, the imperishable nature of your writings will add a great deal to his survival. Happy are they, in my opinion, to whom it is given either to do something worth writing about, or to write something worth reading; most happy, of course, those who do both. With his own books and yours, my uncle will be counted among the latter. It is therefore with great pleasure that I take up, or rather take upon myself the task you have set me.

He was at Misenum in his capacity as commander of the fleet on the 24th of August [sc. in 79 AD], when between 2 and 3 in the afternoon my mother drew his attention to a cloud of unusual size and appearance. He had had a sunbath, then a cold bath, and was reclining after dinner with his books. He called for his shoes and climbed up to where he could get the best view of the phenomenon. The cloud was rising from a mountain-at such a distance we couldn't tell which, but afterwards learned that it was Vesuvius. I can best describe its shape by likening it to a pine tree. It rose into the sky on a very long "trunk" from which spread some "branches." I imagine it had been raised by a sudden blast, which then weakened, leaving the cloud unsupported so that its own weight caused it to spread sideways. Some of the cloud was white, in other parts there were dark patches of dirt and ash. The sight of it made the scientist in my uncle determined to see it from closer at hand.

He ordered a boat made ready. He offered me the opportunity of going along, but I preferred to study-he himself happened to have set me a writing exercise. As he was leaving the house he was brought a letter from Tascius' wife Rectina, who was terrified by the looming danger. Her villa lay at the foot of Vesuvius, and there was no way out except by boat. She begged him to get her away. He changed his plans. The expedition that started out as a quest for knowledge now called for courage. He launched the quadriremes and embarked himself, a source of aid for more people than just Rectina, for that delightful shore was a populous one. He hurried to a place from which others were fleeing, and held his course directly into danger. Was he afraid? It seems not, as he kept up a continuous observation of the various movements and shapes of that evil cloud, dictating what he saw.

Ash was falling onto the ships now, darker and denser the closer they went. Now it was bits of pumice, and rocks that were blackened and burned and shattered by the fire. Now the sea is shoal; debris from the mountain blocks the shore. He paused for a moment wondering whether to turn back as the helmsman urged him. "Fortune helps the brave," he said, "Head for Pomponianus."

3.9 Phases of Eruption of Vesuvius, 79 AD

15 points

(Authored by Emanuela Baer, Shoreline Community College)

At Stabiae, on the other side of the bay formed by the gradually curving shore, Pomponianus had loaded up his ships even before the danger arrived, though it was visible and indeed extremely close, once it intensified. He planned to put out as soon as the contrary wind let up. That very wind carried my uncle right in, and he embraced the frightened man and gave him comfort and courage. In order to lessen the other's fear by showing his own unconcern he asked to be taken to the baths. He bathed and dined, carefree or at least appearing so (which is equally impressive). Meanwhile, broad sheets of flame were lighting up many parts of Vesuvius; their light and brightness were the more vivid for the darkness of the night. To alleviate people's fears my uncle claimed that the flames came from the deserted homes of farmers who had left in a panic with the hearth fires still alight. Then he rested, and gave every indication of actually sleeping; people who passed by his door heard his snores, which were rather resonant since he was a heavy man. The ground outside his room rose so high with the mixture of ash and stones that if he had spent any more time there escape would have been impossible. He got up and came out, restoring himself to Pomponianus and the others who had been unable to sleep. They discussed what to do, whether to remain under cover or to try the open air. The buildings were being rocked by a series of strong tremors, and appeared to have come loose from their foundations and to be sliding this way and that. Outside, however, there was danger from the rocks that were coming down, light and fire-consumed as these bits of pumice were. Weighing the relative dangers they chose the outdoors; in my uncle's case it was a rational decision, others just chose the alternative that frightened them the least.

They tied pillows on top of their heads as protection against the shower of rock. It was daylight now elsewhere in the world, but there the darkness was darker and thicker than any night. But they had torches and other lights. They decided to go down to the shore, to see from close up if anything was possible by sea. But it remained as rough and uncooperative as before. Resting in the shade of a sail he drank once or twice from the cold water he had asked for. Then came an smell of sulfur, announcing the flames, and the flames themselves, sending others into flight but reviving him. Supported by two small slaves he stood up, and immediately collapsed. As I understand it, his breathing was obstructed by the dust-laden air, and his innards, which were never strong and often blocked or upset, simply shut down. When daylight came again 2 days after he died, his body was found untouched, unharmed, in the clothing that he had had on. He looked more asleep than dead.

Meanwhile at Misenum, my mother and I-but this has nothing to do with history, and you only asked for information about his death. I'll stop here then. But I will say one more thing, namely, that I have written out everything that I did at the time and heard while memories were still fresh. You will use the important bits, for it is one thing to write a letter, another to write history, one thing to write to a friend, another to write for the public. Farewell.

3.10 Monitoring of Volcanoes
(Internet Access, Word Processor)   10 points

Name: _____

Due Date: _____

Introduction

Monitoring of potentially active volcanoes is a critical step in preventing volcanic tragedies. The USGS is the organization responsible for monitoring some 150 active or potentially active volcanoes in the United States. Knowing how well volcanoes are monitored is critical to knowing how much warning you and your family might have in a volcanic emergency, and therefore impacts your volcanic risk.

Questions:

1. List and briefly describe 5 ways that volcanoes can be monitored for unrest.

2. **Seismicity** – Earthquakes often provide the earliest warning of volcanic unrest. Swarms of earthquakes often precede most volcanic eruptions. To see a list of seismic stations around volcanoes in the Pacific Northwest, go to http://www.pnsn.org/WEBICORDER/VOLC/welcome.html. Write down the number of seismic stations at each of the 5 potentially active volcanoes in the Cascades. Which ones have had seismic activity in the last 2 months? (Click on "volcanoes" and then on each individual volcano to find a list of recent quakes.)

Volcanoes	# Of Seismic Stations	Quakes in past two months?
Mt. Baker		
Glacier Peak		
Mt. Rainier		
Mt. St. Helens		
Mt. Adams		

3.10 Monitoring of Volcanoes
(Internet Access, Word Processor) 10 points

3. **Ground Movement** - When magma moves up within a volcano, the volcano often changes shape. Outward and upward movement of the ground surface (inflation) usually precedes a volcanic eruption. Geodetic surveys that monitoring ground movement are occasionally completed at some Washington volcanoes. Go to http://vulcan.wr.usgs.gov/Projects/Deformation/cvo_networks.html. Which Washington state volcanoes are monitored for deformation and when was the last time that they were checked for changes?

4. **Hydrologic Regime** - Measuring changes in groundwater temperatures or level, lake levels, and rates of stream flow can help scientists evaluate the role of groundwater in generating volcanic eruptions and the potential hazards of lahars (mudflows). Unfortunately, only Mount St. Helens is currently monitored in real time this way. Go to: http://vulcan.wr.usgs.gov/Monitoring/framework.html, click on "Mount St. Helens Hydrologic Real-Time Monitoring". Look through the list of monitoring stations. What type of data is collected?

5. Summarize the current monitoring and activity levels for each of Washington's volcanoes.

6. Do you think that your volcanoes are adequately monitored? What would you do to change the monitoring of our volcanoes? Which one/ones are more in need of increased surveillance?

7. Type your responses to questions 5 and 6 and submit them with correctly formatted bibliographic references for any sources you used.

4.1 Angle of Repose

(Materials provided in class)   10 points

In this activity you will measure the maximum slope at which grains are stable (angle of repose). You will explore how different properties of the sediment influence slope stability and lead to different slope failures (mass movements).

Materials

- Pan or tray
- Sediment samples: dry sand, damp sand, angular gravel, rounded gravel
- protractor
- small scoop
- container of water

Instructions (Read these carefully or you will have to redo your responses!)

1. Slowly pour a stream of dry sand into the center of your pan or tray. Avoid disturbing your pile and carefully measure the maximum angle, or steepest slope, of the dry sand. This slope is called the angle of repose. Record your answer on the data sheet (1).
2. Add a few grains (a pinch) of dry sand to the top of your pile and see how the grains move. Repeat, observing the movement again. Describe the downward movement of the dry sand on the data sheet (2). Note whether the sand grains move individually or in large groups.
3. Put all your dry sand back into its container. Sweep your tray clean (use the garbage cans).
4. Repeat the experiment using damp sand. Place a pile of damp sand in the center of your pan or tray. You will probably have to scoop the sand instead of pour it. You can gently pack the pile together, but use only gentle pressure. What is the maximum angle of the damp sand? (3)
5. Add a pinch of the damp sand to the top of the pile and see how the grains move. Repeat, observing the movement again. Describe the downward movement of the damp sand on the data sheet (4). Note whether the grains move individually or in large groups.
6. With the cup, _slowly_ pour water onto the pile of damp sand and observe what happens to the sand pile as the sand becomes saturated. Can the saturated sand maintain its angle of repose? (5).
7. Again, clean up your mess and answer questions (6) – (9).
8. Slowly pour the rounded gravel into the center of your tray or pan. Make sure your pile is large enough to get the true angle of repose! Carefully measure the angle of repose, without disturbing the pile. Record angle (10). Pour the rounded gravel back into the container.
9. Repeat the experiment using the angular gravel and record the results (11). Answer questions (12 and 13).
10. Put all materials away. Clean up your area.

4.1 Angle of Repose

(Materials provided in class)

10 points

4.1 Angle of Repose
(Materials provided in class)   10 points

DATA SHEET   Name: _____

                              Group Members: _____

                                            _____

                                            _____

                                            _____

1.   Dry sand angle of repose is _____ degrees.

2.   Dry sand movement:

     _____

     _____

3.   Damp sand angle of repose is _____degrees.

4.   Damp sand movement:

     _____

     _____

5.   Can the **saturated** sand maintain a steep angle of repose? _____

6.   Which sediment condition – dry or damp – permits steeper slope angles? _____

7.   Which sediment condition – dry or damp – lends itself to the most dramatic,
     quickest, and therefore most dangerous style of slope failure? _____

8   Explain why you think the damp sand can maintain a higher angle of repose than
     the dry sand.   _____

     _____

     _____

9.   Suppose a house is built on a slope made of sediment. The slope is slightly
     greater than the angle of repose. Is this house at risk from mass movement?

                                            _____

10. Rounded gravel angle of repose is _____ degrees.

11. Angular gravel angle of repose is _____ degrees.

12. Compare the angle of repose for the dry sand and the dry gravel. What impact
     does grain size have on the angle of repose?

     _____

     _____

     _____

13. Compare the angle of repose for the rounded gravel and the angular gravel. What
     impact does grain shape have on the angle of repose?

     _____

     _____

     _____

4.1 Angle of Repose
(Materials provided in class)

10 points

Use the results of your experiments and the graphs below to answer the following questions. (If you are completing this at home, please type your answer as it will be useful in your paper. Otherwise, please use a separate sheet of paper.)

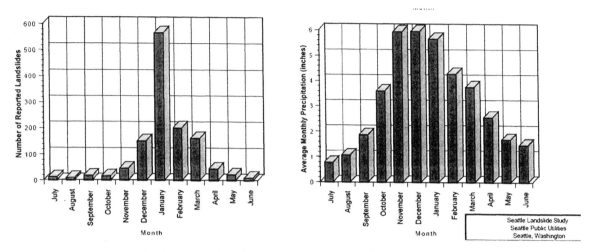

Source of data: City of Seattle, 1999. Seattle Landslide Study
http://www.seattle.gov/DPD/Landslide/Study/default.asp accessed August, 2005.

14. In the Seattle area, which four months experience the greatest amount of rainfall?

15. In the Seattle area, which four months experience the highest number of landslides?

16. Thinking about the results of your experiments (especially in Step 6), explain why the peak landslide activity occurs later than the peak rainfall activity.

4.1 Angle of Repose
(Materials provided in class)   10 points

The USGS has recently developed and tested a model that establishes a precipitation threshold for the Puget Sound Area as it relates to landslides. In this area, landslides tend to occur if the precipitation amount for a 15 day period exceeds a specified level, and that 15 day period is followed by 3 days of rain at certain levels. The graph below shows the Precipitation Threshold for Anticipating the Occurrence of Landslides.

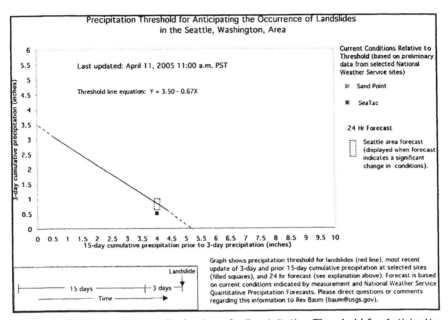

Source: Chleborad, A.F., (2003) <u>Preliminary Evaluation of a Precipitation Threshold for Anticipating the Occurrence of Landslides in the Seattle, Washington, Area.</u> U.S. Geologic Survey OFR 03-463. Available online: http://pubs.usgs.gov/of/2003/ofr-03-463/ofr-03-0463.html.

17. Place and label a point on the graph for the following days:

Date	15 day cumulative precipitation	3 day cumulative precipitation	Landslides expected? (Yes, no, maybe)
1/5/56	4.73	3.38	
1/6/96	1.27	0.55	
1/7/96	1.33	1.27	
2/11/96	4.94	0.07	
2/21/96	5.87	0.56	

18. Explain why the 15 day and 3 day cumulative precipitation would be critical in determining when landslides might occur.

133

4.1 Angle of Repose
(Materials provided in class)   10 points

19. Why might landslides not occur after even several inches of rain in 3 days if the amount of rain in the preceding 2 weeks has been low?

20. Why would landslides not occur even if there had been a large amount of rain in the previous 2 weeks, but little rain in the last 3 days?

4.2 Forces in Mass Movement
(Ruler, Calculator, Protractor)   10 points

Name _____
Due Date _____

Forces are usually represented by vectors; graphically we draw these as arrows. The length of the arrow represents the amount of force, and the while the direction of the arrow is the direction of the force. So a gravitational force is represented by an arrow downward, and the larger the arrow the larger the mass of the object.

A small mass object has a smaller arrow. In this case the arrow on the second object is twice as long showing the object is twice as heavy.

Two vectors are equal if they have the same magnitude and direction, regardless of whether they have the same starting points.

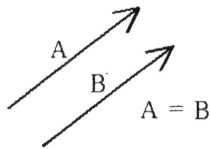

Vectors are added graphically by placing the initial point of one vector on the final point of the other. This is sometimes referred to as the "Tip-to-Tail" method.

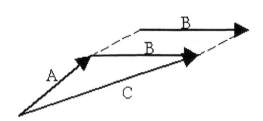

A+B = C
Notice that you could also have moved the tail of A to be at the head of B and still get the same result

135

4.2 Forces in Mass Movement
(Ruler, Calculator, Protractor)

10 points

NOTE: *The length of A plus the length of B does NOT equal the length of C... Vectors are not just lengths, but lengths AND directions!*

Just as you can add vectors, you can divide them into multiple forces. This is called "resolving a vector." In the above example, you can divide C into two components A and B.

When dealing with landslides, we are interested in the interplay between two forces, friction (F), which works against objects moving downslope, and the pull of gravity down-slope (G_d). To find G_d we must resolve the gravitational force (G) into two components, one perpendicular to the slope (G_p) and one down the slope (G_d).

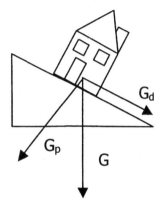

Part I – Questions relating to vectors:

1. Which of the following vectors are equal?

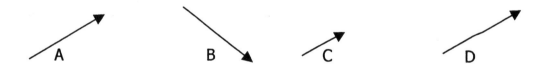

2. Draw the vector created by adding vectors A and B.

3. What is the length of vector A? What is the length of vector B? What is the length of vector C? (Use a ruler and measure in millimeters.)

136

4.2 Forces in Mass Movement
(Ruler, Calculator, Protractor)   10 points

4. What vector must be added to vector A to get vector C? Draw it in. What is its length?

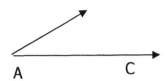

Part II - Questions relating to landsliding:

1. Mark the following diagrams of houses on slopes as "stable" "unstable" or "just about to slide".

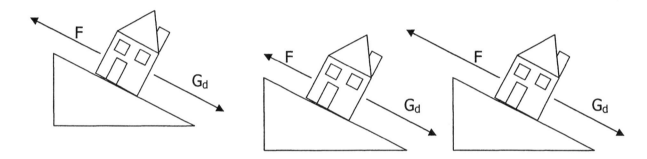

2. On the following diagrams, resolve the gravitational force (G) into a downslope (G_d) and a perpendicular (G_p) force. In the first example, the dashed lines show the direction of the G_d and G_p forces. You just need to draw the two vectors so that they add up to G. In the others, you will need to make sure the directions and lengths are correct. Please label the vectors you draw.

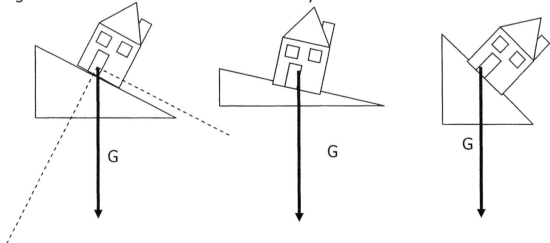

137

4.2 Forces in Mass Movement
(Ruler, Calculator, Protractor)   10 points

3. The G vectors in the previous example are all the same length, 40 mm.

   If the house weighs 40,000 pounds, then each millimeter of length of the vector in this diagram equals how many pounds? _____

   How long would the G vector be if the house was 60,000 pounds? _____

   Measure G_d in each diagram in problem 2. How many pounds of force are pulling the house downhill in each case?

4. A boring (drilling) into the ground examines the friction on the slope below the house and finds that the friction is 21,000 lbs. Draw this force into the pictures above, then label each figure as stable, unstable or just about to slide.

5. A developer buys a plot of land with a 25° slope. The friction in the slope is measured at 60,000 kilograms. Draw a diagram of this and figure out what the maximum weight a house can be on this property. (BONUS! Show how you can calculate this using trigonometry instead of graphically.)

4.3 Examining Slope Stability

10 points

Name _____

Due Date _____

In this activity you are provided with cross-sections of hills exposing different geologic conditions. You will be asked to evaluate the stability of each hill and, if unstable, suggest a landslide prevention or mitigation technique that may be appropriate for that situation.

The following is a key to the rock type symbols used in the cross-sections:

Symbol	Rock Type	Notes on properties:
	Granite	A coarse-grained igneous rock (crystalline), with high strength and low porosity and permeability unless highly fractured.
	Sandstone	A sedimentary rock made of sand-sized grains held together by natural cements. Sandstones can have high strength, moderate porosity/ permeability.
	Shale	A weak sedimentary rock made of compacted mud (clay/silt). The rock has low porosity and permeability and can break apart into thin sheets.
	Limestone	A sedimentary rock made of calcite, a mineral that dissolves in weak acidic water. Limestone has moderate strength.
	Consolidated sediments	Compacted mixtures of sand, gravel, silt and clay, like glacial till.
	Unconsolidated Sediments	Sorted sand and gravel deposits, usually compacted, but not cemented.

Instructions
Examine the cross-sections on the following pages. For each cross section, you should write a short paragraph considering each of the following:

- Is the slope stable, somewhat stable, somewhat unstable or very unstable?
- Describe why you think it is stable or unstable based on the geologic conditions.
- If the slope is unstable, how might the slope fail? Do you think there will be a slump, translational slide, fall, or flow?
- Considering the mitigation and prevention techniques described in class. What technique might work to stabilize the slope shown in the cross-section?

In all cross-sections, you should consider an average (not excessive) amount of rainfall.

4.3 Examining Slope Stability

10 points

4.3 Examining Slope Stability

10 points

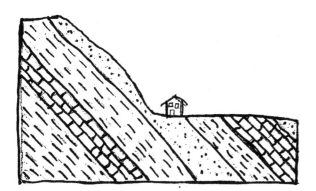

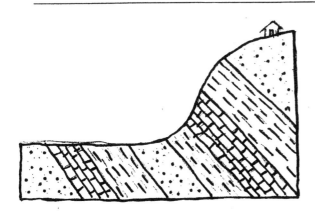

4.3 Examining Slope Stability

10 points

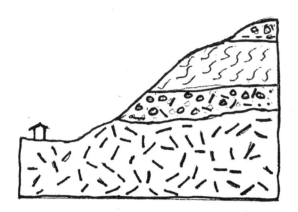

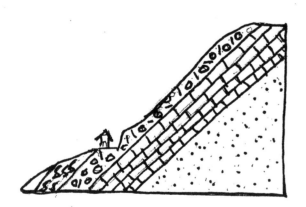

4.3 Examining Slope Stability

10 points

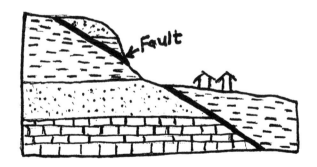

4.4 The Liquid Earth
(Internet Access, Word Processor)   15 points

Name _____

Due Date _____

Copies of the article "The Liquid Earth", by Brenda Bell will be distributed by your instructor. This article was published in the Atlantic Monthly in January 1999.

Please read the article and answer the following questions on a separate piece of paper. Answers will be graded on accuracy, completeness, and clarity. As always, responses should be typed or in a neat enough form me to read.

Your responses are due at the beginning of the class period on the due date.

1. Each year the United States government supports the U.S. Geological Survey to do hazard research. How does the spending on landslide hazards compare with the spending on earthquake hazards? Explain why you agree or disagree with this allocation.

2. In 1982 an episode of mass wasting events took place in the area around San Francisco, California. What triggered these events, how many debris flows occurred, and what was the cost in terms of human casualties and damage?

3. Clearly explain what is meant by "hard" and "soft" solutions to landslide problems? Your explanation should include several examples of each.

4. One area where geologists attempted to study a potential landslide before it occurred was at Mettman Ridge, near Coos Bay Oregon. Describe this natural laboratory, and the experiments that were being conducted there. What ultimately happened to the site?

5. After having read this article, what do you think would be two (2) viable solutions to the landslide problems in the area around the Puget Sound region? Which of the two (2) solutions you suggest would be best, in terms of future development?

4.5 Landslides in the Puget Sound
(Internet Access, Word Processor)   15 points

Name _____

Due Date _____

In this activity, you will visit two websites that will help you to learn about the glacial geology of the Puget Sound area and how it contributes to landslide hazards in this area.

NOTE: The information you obtain from this assignment <u>must</u> be incorporated into your Landslide Hazard and Risk Report in which you describe the hazards of landslides in the Puget Sound.

You should first visit the following website and answer the following questions:

Department of Ecology, *About Slides on Puget Sound*,
[http://www.ecy.wa.gov/programs/sea/landslides/about/about.html]

1. What are the six factors that contribute to landslide hazards in the Puget Sound?

2. When do most Puget Sound landslides occur and why?

3. Identify three ways in which human actions contribute to landslides.

4.5 Landslides in the Puget Sound
(Internet Access, Word Processor)   15 points

4. Click on the link to "Geology" at the web site and examine the photograph and cross-section describing the common soil layering sequence of hills in the Puget Sound region. Draw a simple cross-section showing a typical hillside and label and describe the three layers found there.

Next, visit the following website, read the Introduction and the section "Seattle Area Stratigraphy" and answer the questions below:

Department of Natural Resources, March 1997, Puget Sound Bluffs: The Where, Why and when of Landslides Following the Holiday 1996/97 Storms: Washington Geology, vol. 25, no. 1. [http://www.dnr.wa.gov/geology/pugetls.htm]

5. After reading the introduction, click on Figure 1. This map shows the shows the landslide critical areas (shaded grey) in the City of Seattle. Did most of the landslides during the 1996/97 storms occur in these known landslide hazard areas?

6. Read the section "Seattle Area Stratigraphy" and examine Figure 20. What is the name of the glacial till that appears at the top of many of the hills in this area? (Label your cross-section in question 4 with the name). When was this material deposited by glaciers?

7. What is the name of the unit underlying the glacial till in most areas of Seattle? How was this material deposited? (Label your cross-section in question 4 with the name.)

148

4.5 Landslides in the Puget Sound
(Internet Access, Word Processor)   15 points

8. What is the name of the bottom layer of sediments found in your cross-section? Where/how was this material deposited? (Again, label your cross-section in question 4 with the name.)

9. Examine Figure 21 and read the figure caption. In your own words, describe how the geologic layering contributes to landslides in this area.

10. On a separate sheet of paper, type a paragraph or two describing the glacial geology of the Puget Sound area and another paragraph or two on how/why it contributes to landslides in the Puget Sound area. Be sure to make the appropriate citations and include a bibliography.

4.6 Mass Movement Hazards Around Your Home
(Internet Access, Camera, Word Processor)   20 points

Name _____

Due Date _____

This assignment will have you examine a variety of sources to determine if your residence or neighborhood is at risk for landsliding or other mass movements. It involves both library research and looking at your home, so it cannot be done at the last minute. Please plan accordingly.

Part 1: The Slope You Live On!

You will be using topographic maps to determine the grade of the slope you live on.   In class, select the topographic map that includes the area where your home is. Locate this location on the map and answer the following questions.

1. What is the name of the map you are using?

2. Look at the bottom and find the contour interval. What is the contour interval for the map?

3. Find the location of your house on the map, but do not mark on it PLEASE. Take a ruler, and, without marking on the maps, measure the distance between the two contours nearest your house. You will either measure in inches if you map is in feet or centimeters if your map is in meters. Use the scale at the bottom of the map to determine the number of feet (or meters) that this distance is. You will need to be very careful and exact on your measurements.

    Measurement on the ruler: Inches _____ or Centimeters _____

    Distance (using the map scales): Feet _____ or Meters _____

4. To find the slope of your area, take the contour interval and divide by the number above. This number should be less than 1, so multiply by 100 to get a slope in percent.

    EXAMPLE:
    Contour interval in Des Moines: 25 ft (read from the bottom of the map)
    Distance across the two contours nearest my house: 0.1 inches (measured from the map)

    Using the scale at the bottom, I see that 0.1 inches is 200 ft.
    My slope is 25/200 = .125 or 12.5%

    The slope at my house is _____.

151

4.6 Mass Movement Hazards Around Your Home
(Internet Access, Camera, Word Processor)   20 points

5. How far away is the nearest stream or body of water from your residence? What is its name?

6. Where is the nearest steep slope from your residence? How far away is it?

7. Write the bibliographic reference for this topographic map.

Part 2: Hazards Maps

Your assignment is to find out if you live in an erosion, landsliding or steep slope area. This can be done by looking at your local city's Sensitive Areas maps. You can find these in a variety of ways:
1. Find the maps at the Highline Library – we have gathered quite a collection!
2. Talk to your instructor – he or she might have a very relevant map for you to look at.
3. Look on-line. Many (although not all) cities have their maps on line.
4. Look in your local public library. These maps will be in the Local Documents areas.
5. Go to your local city government planning/building office.
6. Finally, if all else fails, or if you are in an unincorporated area, look at the King County Sensitive Areas maps, available at the Highline Library.

1. What city do you live in?

2. Do you live in any sensitive areas relating to mass wasting, according to the maps? (This includes steep slopes, erosion, as well as landsliding.)

3. According to the map(s) you looked at, where is the closest area that is at high risk for mass movement? How far away is this from your residence?

4.6 Mass Movement Hazards Around Your Home
(Internet Access, Camera, Word Processor)   20 points

4. Write a bibliographic reference for all the maps you examined.

5. Do you live within _ mile of the coast? _____. If the answer is yes, please look at the coastal Atlas of Washington at http://www.ecy.wa.gov/programs/sea/SMA/atlas_home.html and determine your slope stability. Record what you find here.

Part 3: Looking for Mass Wasting at Your Residence

Now, you will look for signs of mass wasting in your neighborhood. To begin this, you should consult your book, notes and the internet to answer questions 1-3, to find what you will be searching for.

1. What are 5 signs that an area is undergoing creep?

2. What are 3 common ways that people engineer solutions to mass wasting?

3. Go to http://www.ecy.wa.gov/programs/sea/landslides/signs/signs.html and peruse the signs of movement list. What are the 4 broad categories of evidence that you might be looking at a place with mass movement?

4.6 Mass Movement Hazards Around Your Home
(Internet Access, Camera, Word Processor)   20 points

For the next four questions, please <u>type</u> your responses on a separate piece of paper. Use the pieces of evidence listed at the above website to <u>specifically</u> determine if your area has signs of movement. For instance, "there are no signs of movement" is not sufficient. Each piece of evidence you can use needs to be addressed.

4.   Is there any evidence of creep or slow mass movement at your residence or in your neighborhood? Please describe and take photographs of any evidence you find and submit with this report. If you do not find evidence, please describe what evidence you *did not* find that would lead you to believe that there is no slow mass movement in your area.

5.   Is there any evidence of faster mass movements at your residence or in your neighborhood? Please describe and take photographs of any evidence you find and submit with this report. If you do not find evidence, please describe what evidence you *did not* find that would lead you to believe that there is no faster mass movement in your area.

6.   Did you find any evidence of engineering solutions to mass wasting either at your residence or in your neighborhood? Please describe and take photographs of any evidence you find and submit with this report.

7.   At the end of your report, please include a correctly formatted bibliographic citation for the Department of Ecology website you used.

8.   Finally, go to the nearest steep area (the one you named from the topographic map). Describe what you find, including any evidence of mass wasting of any type, and any human response to it. Once again, please take photographs and submit with this homework.

4.7 Video: Mass Wasting

10 points

Name _____

Due Date _____

This video is part of the series "Earth Revealed", Program 16. It is available on reserve at the Circulation Desk in the school library.

1.   Where did a massive mudflow take the life of 22,000 people in 1985?

2.   What's the main force that causes mass wasting?

3.   What are 3 other factors that contribute to mass wasting?

4.   What is the slowest type of mass wasting?

5.   What causes creep?

6.   What are some pieces of evidence for creep?

7.   What type of mass movement happened at Point Fermin, CA between 1929 and 1940?

4.7 Video: Mass Wasting

10 points

8. What are some mitigation strategies that have been used to correct an unstable slope?

9. Why are debris flows and mudflows so destructive?

10. How can debris or mudflow be controlled?

11. What are the most massive and potentially destructive types of mass wasting?

12. What type of field surveys can be done to evaluate the stability of a slope?

5.1 Flood Frequency in Seattle Area

(ruler, pencil)   10 points

(Adapted from: Duncan, F., McKenzie, G.D., and Utgard, R.O., 1999, Investigations in Environmental Geology, Prentice-Hall, Inc., pp. 303. Data modified from USGS Fact Sheet 229-96, "The 100-Year Flood".)

Introduction

The population in the Puget Sound is growing rapidly and humans have made many changes to rivers and drainages. This activity investigates recurrence intervals, 100-year floods, and changing flood frequencies for two watersheds in the state of Washington.

Gauges placed along a stream monitor the discharge (volume of water in the stream) at several locations. The USGS and other organizations collect the data from gauging stations and use it to determine the frequency of flooding along the stream. Estimates of **flood frequency** are more accurate with a long record (many years) of discharge records. The flood frequency is typically expressed as a **recurrence interval** (the average number of years between the occurrence of two floods with the same level of discharge). To calculate the recurrence interval, the peak annual flood of the stream is used. The **peak annual flood** is the highest discharge recorded for the year at a gauging station.

The formula for determining the recurrence interval (T, in years) for a flood of a given discharge is:

$$T = \frac{(n+1)}{m}$$

where: n = the number of years of record
                m = the rank or order of the annual flood discharges from the greatest
                (1) to the smallest for the number of years of record.

From the peak annual flood data, the recurrence interval is calculated and plotted on a **flood frequency graph**. The line on a flood frequency graph allows geologists to estimate the average number of years that will elapse until a flood of a particular magnitude reoccurs. Flood frequency graphs are used in flood prediction.

If peak annual flood data has been collected over a period of many years, there is a high probability that floods of all sizes, large and small, have been recorded. We use the flood frequency graph to determine the discharges associated with a 20-year, 50-year or 100-year floods. Obviously, the estimate of these floods is more reliable if they are based on a many years of data rather than if they are based on relatively few years.

The 100-year flood, as determined by this type of flood frequency analysis, serves as a legal definition of areas that are likely to be flooded. If someone chooses to purchase a home in the 100-year floodplain, they must obtain flood insurance provided by the federal government.

5.1 Flood Frequency in Seattle Area
(ruler, pencil)   10 points

Instructions

Divide into groups of 4 students. Two people in your group will be working with Mercer Creek (Data Sets 1 and 2). Two people in your group will be working with the Green River (Data Sets 1 and 2). Each data set spans 11 years of record along the river. Use your assigned data set to estimate the likely discharge for a 100-year flood for each 11-year period. You should follow this procedure:

1.   For each data set you have been assigned, rank the peak flood discharge in order of magnitude, starting with 1 for the largest and ending with 11 for the smallest. Write these results in the "Rank" column of the table.

2.   Use the formula (T= (n+1)/m) to determine the recurrence interval of each of the 11 floods in each of the two data sets. The results should be recorded in the "Recurrence Interval" column of the table.

3.   Examine the graph paper provided in this activity. The Recurrence Interval is plotted along the horizontal axis. The graph begins at 1.01 years and increases to the right. Determine the value, in years, of the lower right corner of the graph. Write the number below the right edge of the graph.

4.   You will be plotting the peak flood discharge for your assigned stream on the vertical axis of the graph. To do this you need to select an appropriate vertical scale for your discharge data by examining the value of the highest and lowest discharges found in your two data sets. For instance, if the low discharge is 500 ft^3/s and the high discharge is 4000 ft^3/s, then your range of discharge is 3,500 ft^3/s. *Choose a vertical scale so that the numbers you plot from your data fill about one-half of the length of the vertical axis.* Label the vertical axis by writing the appropriate numbers for your discharge along the left edge of the graph paper.

5.   For your assigned stream, plot the discharge and recurrence interval for each of the 11 floods from Data Set 1. Using a ruler, draw a best-fit straight line through the data points. *(If you do not know how to draw a best-fit line, ask your instructor.)* The line should be extended all the way to the right side edge of the graph.

6.   Now plot the discharge and recurrence interval for each of the 11 floods from Data Set 2 on the same graph paper. Using a ruler, draw a best-fit straight line for this data. The line should be extended all the way to the right side edge of the graph.

5.1 Flood Frequency in Seattle Area
(ruler, pencil)   10 points

Mercer Creek — Data Set 1				Mercer Creek — Data Set 2			
Year	Peak Flood Discharge	Rank (1 = greatest)	Recurrence interval	Year	Peak Flood Discharge	Rank (1 = greatest)	Recurrence Interval
1957	180			1979	518		
1958	238			1980	414		
1959	220			1981	670		
1960	210			1982	612		
1961	192			1983	404		
1962	168			1984	353		
1963	150			1985	832		
1964	224			1986	504		
1965	193			1987	331		
1966	187			1988	228		
1967	254			1989	664		

Green River — Data Set 1				Green River — Data Set 2			
Year	Peak Flood Discharge	Rank (1 = greatest)	Recurrence interval	Year	Peak Flood Discharge	Rank(1 = greatest)	Recurrence Interval
1941	9310			1976	4490		
1942	10900			1977	9920		
1943	12900			1978	6450		
1944	13600			1979	8730		
1945	12800			1980	5200		
1946	22000			1981	9300		
1947	9990			1982	10800		
1948	6420			1983	9140		
1949	9810			1984	10900		
1950	11800			1985	7030		
1951	18400			1986	11800		

5.1 Flood Frequency in Seattle Area

(ruler, pencil)
10 points

Recurrence Interval (in years)

1.01 1.1 1.2 1.3 1.4 1.5 2 3 4 5 6 7 8 9 10 20 30 40 50

5.1 Flood Frequency in Seattle Area

(ruler, pencil)   10 points

5.1 Flood Frequency in Seattle Area
(ruler, pencil)   10 points

Worksheet

1. Based on your data, what is the predicted discharge for a 100-year flood? To find this information, you must read the value from your graph where it intersects the 100 yr recurrence interval line. Send someone for your group to record this information on the board, so that others in the class can record it.

Data Sets	Predicted discharge for a 100-year flood
Mercer Creek – Data Set 1 (1957-1967)	
Mercer Creek – Data Set 2 (1979-1989)	
Green River – Data Set 1 (1941-1951)	
Green River – Data Set 2 (1976-1986)	

2. How do your predictions for the river you have been assigned compare to each other. Remember you are comparing **for the same river**, Data Set 1 and 2. Describe it in words.

3. Suggest possible human activities in the watershed that could have caused the differences in predicted floods that result from the two sets of data for your river.

4. How do the data from your stream compare with data from the other stream? What human activities can you suggest for the changes in flood predictions the other group discovered?

5.1 Flood Frequency in Seattle Area
(ruler, pencil)
10 points

5. Based on the flood predictions for all four data sets, what does the contrast in predicted flood discharges imply about the usefulness of the 100-year flood as a legal designation for these two streams?

6. What information do you need to know if you are about to buy a house that is located adjacent to, but just outside the 100-year floodplain?

It is possible to calculate the probability (or chance) that the annual maximum flood will equal or exceed a given discharge within any single year. This is called the annual probability of exceedence, P, and it is the reciprocal of T (the recurrence interval). Written as a formula:

$$P = \frac{1}{T}$$

Use the formula to calculate the following:

7. What is the probability in any given year that the stream discharge will exceed the discharge of the 100-year flood recurrence interval?

8. What is the probability in any given year that the stream discharge will exceed the discharge of the 10-year flood recurrence interval?

5.2 Flooding in Your Neighborhood
(FEMA Maps)   10 points

Name _____

Due Date _____

In this assignment, you will look at maps compiled by the Federal Emergency Management Association (FEMA) that shows flooding hazards. The first step is to find the FEMA Flood Map for your community/area. These maps are available from your instructor or at the School Library. Once you have located the appropriate map, answer the following questions:

1.   What is the title of the map that shows the location of your residence?

2.   When year was the map made? The date should be located somewhere on the map.

3.   Look at the map scale. How far is one inch on this map? (In terms of the actual distance.)

4.   Do you live in a flood plain? If the answer is yes, which one? For which body of water? If the answer is no, how close is the nearest Special Flood Hazard Area? The closest 100- and 500-year flood zones?

5.   What flood insurance risk zones exist in your community? What is in those high-risk zones (i.e., housing development, retail shops, etc.)?

5.3 Video: Running Water

10 points

Name _____

Due Date _____

(From the video "Running water, erosion and deposition" - The Earth Revealed Series, Program 19; available on reserve at the Circulation Desk in the school library.)

1. Where is the Nile River?

2. Is river flow greater or less than the precipitation in a drainage basin?

3. Name two landscapes that rivers form.

4. What substrate causes the greatest resistance to flow?

5. What is the discharge of a river?

6. What units is discharge measured in?

7. What are 3 different methods by which sediment can be picked up (eroded) in a river?

8. What are 3 different processes/ways in which sediment can be transported in a river?

9. What is a "bar" in a river?

10. What two types of rivers are bars found in?

11. Where are point bars found?

5.3 Video: Running Water

10 points

12. Where are cut banks found?

13. How often does a typical river flood (come out of its banks)?

14. Why are flood plains good for agriculture?

15. Where do levees form?

16. What effect do dams have on sediment?

17. How long is the Mississippi?

18. What happens to the sediment at Red Eye crossing?

19. Name one idea of how to keep the crossing open to commercial navigation.

20. Were dikes installed at Red Eye Crossing? Why or Why not?

21. On average how many people are killed in flood in the US?

5.4 Video: Running Water – Landscape Evolution

10 points

Name _____

Due Date _____

(From the video "Running water, landscape evolution" - The Earth Revealed Series, Program 20; available on reserve at the Circulation Desk in the school library.)

1.   How is running water different than earthquakes and volcanoes?

2.   How long did it take the Colorado River to carve the Grand Canyon?

3.   What two processes cause the formation of river valleys?

4.   What is the "great base level" below which rivers cannot erode?

5.   In a recently uplifted area, how do streams expend their energy?

6.   As the downcutting finishes, how do streams expend their energy?

7.   What causes the differing slopes in the Grand Canyon?

8.   What is the name of the drainage pattern formed by streams that develop on homogeneous or flat lying rocks?

9.   What kind of drainage pattern formed by streams that develop on intersecting fractures or faults?

5.4 Video: Running Water – Landscape Evolution

10 points

10. What landform is caused by rejuvenation?

11. What is the thalweg?

12. Other than valleys, what is another major landform created by rivers?

13. Why do deltas form?

14. What is a distributary?

15. How many tons of sediment are added to the delta of the Mississippi every day?

16. Which Mississippi delta cities depend on the Mississippi?

17. Why is the Achafalaya River likely to capture the Mississippi?

18. When were the river control structures on the Old River approved?

19. What control structures are on the Old River?

20. What is the goal of these control structures?

21. What year was the great flood that nearly destroyed the Old river control structure?

22. How much did the auxiliary structure cost?

5.4 Video: Running Water – Landscape Evolution

10 points

23. As a result of loss of sediment, the Achafalaya scours its bed. What does this do the elevation of the river?

24. As a result of deposition of sediment, what is happening to the elevation of the Mississippi?

25. The shape of the Earth's surface is dominated by the constant struggle between what two forces?

6.1 Reference Style Guide for Paper/Assignments

All references mentioned ("cited") in the paper – no matter where, must be listed in a "References Cited" section. Only references cited in the paper are to be listed. At the end of the text, list references alphabetically by author's surname. Do not abbreviate journal titles or book publishers in references. Please indicate the city of publication for books. For references that do not match any of the examples given here, include *all information that would help a reader locate the reference*.

Correct citation format for your final paper is as follows. These examples are from the Geologic Society of America.

WWW SOURCES:
In general, please use:

> Author's name (last name, first and any middle initials). Date of Internet publication. Document title. <URL> or other retrieval information. Date of access.

Some examples:

Online Book (may be the electronic text of part orf all of a printed book, or a book-length document available only on the Internet).

> Bryant P. 1999 Aug 28. Biodiversity and conservation. <http://darwin.bio.uci.edu/~sustain/bio65/index.html>. Accessed 1999 Oct

Article in an electronic journal (ejournal)

> Browning T. 1997. Embedded visuals: student design in Web spaces. Kairos: A Journal for Teachers of Writing in Webbed Environments 3(1). <http://english.ttu.edu/kairos/2.1/features/browning/bridge.html>. Accessed 1997 Oct 21.

Newspaper Article

> Azar B, Martin S. 1999 Oct. APA's Council of Representatives endorses new standards for testing, high school psychology. APA Monitor. <http://www.apa.org/monitor/tools.html>. Accessed 1999 Oct 7.

Government publication with author
> Bush G. 1989 Apr 12. Principles of ethical conduct for government officers and employees. Executive Order 12674. Part 1. <http://www.usoge.gov/exorders/eo12674.html>. Accessed 1997 Nov 18.

Government publication without author (use sponsoring organization instead)
> USGS, 2002 Apr 23, Mount Adams, Washington: Cascade Volcanoes Summary, <http://vulcan.wr.usgs.gov/Volcanoes/Adams/summary_mount_adams.html>. Accessed 2004 Aug 16.

6.1   Reference Style Guide for Paper/Assignments

FILM, VIDEOTAPE OR DVD:

Lienau, M. (Producer), 2003. Fire Mountains of the West: The Cascade Volcanoes [DVD video], Global Net Productions, Inc.

PRINT SOURCES:

Journals

Arias, O., and Denyer, P., 1991, Estructura geológica de la región comprendida en las hojas topográficas Abras, Caraigres, Candelaria y Río Grande, Costa Rica: Revista Geológica de América Central, no. 12, p. 61–74. [Use correct diacritical marks for all non-English languages.]

Doglioni, C., 1994, Foredeeps versus subduction zones: Geology, v. 22, p. 271–274.

Leigh, D.S., 1994, Roxana silt of the Upper Mississippi Valley: Lithology, source, and paleoenvironment: Geological Society of America Bulletin, v. 106, p. 430–442.

Walter, L.M., Bischof, S.A., Patterson, W.P., and Lyons, T.L., 1993, Dissolution and recrystallization in modern shelf carbonates: Evidence from pore water and solid phase chemistry: Royal Society of London Philosophical Transactions, ser. A, v. 344, p. 27–36.

Maps

Abrams, G.A., 1993, Complete Bouguer gravity anomaly map of the State of Colorado: U.S. Geological Survey Miscellaneous Field Studies Map MF-2236, scale 1:500 000, 1 sheet.

Bayley, R.W., and Muehlberger, W.R., compilers, 1968, Basement rock map of the United States, exclusive of Alaska and Hawaii: U.S. Geological Survey, scale 1:2 500 000, 2 sheets.

Ernst, W.G., 1993, Geology of the Pacheco Pass quadrangle, central California Coast Ranges: Geological Society of America Map and Chart Series MCH078, scale 1:24 000, 1 sheet, 12 p. text.

WA DNR, 2002, Tsunami Inundation Map of Port Townsend, Washington Area, Division of Geology and Earth Resources, Open File Report 2002-2, scale 1:24 000, 1 sheet.

Books

Marshak, S., 2005, Earth: Portrait of a Planet (2nd ed.), New York: W.W. Norton & Company, Inc.

Decker, R. and Decker, B., 1998, Volcanoes (3rd ed.), New York: W. H. Freeman & Company, Inc.

Abstracts

Baker, V.R., 1991, Ancient Oceans on Mars: American Astronomical Society Division of Planetary Sciences, 23rd Annual Meeting, Palo Alto, California, Abstracts, p. 99.

Fitzgerald, P.G., 1989, Uplift and formation of Transantarctic Mountains: Applications of apatite fission track analysis to tectonic problems: International Geological Congress, 28th, Washington, D.C., Abstracts, v. 1, p. 491.

LeMasurier, W.E., and Landis, C.A., 1991, Plume related uplift measured by fault displacement of the West Antarctic erosion surface, Marie Byrd Land [abs.]: Eos (Transactions, American Geophysical Union), v. 72, p. 501.

Sammis, C.G., 1993, Relating fault stability to fault zone structure: Geological Society of America Abstracts with Programs, v. 25, no. 6, p. A115–A116.

6.1   Reference Style Guide for Paper/Assignments

Paper in a Government or University Serial Publication
Hay, R.L., 1963, Stratigraphy and zeolitic diagenesis of the John Day Formation of Oregon: University of California Publications in Geological Sciences, v. 42, p. 199–262.

Smith, D.C., Fox, C., Craig, B., and Bridges, A.E., 1989, A contribution to the earthquake history of Maine, *in* Anderson, W.A., and Borns, H.W., Jr., eds., Neotectonics of Maine: Maine Geological Survey Bulletin 40, p. 139–148.
Yager, R.M., 1993, Estimation of hydraulic conductivity of a riverbed and aquifer system on the Susquehanna River in Broome County, New York: U.S. Geological Survey Water-Supply Paper 2387, 49 p.

Paper in a Multiauthor Volume
Carpenter, F.M., 1992, Superclass Hexapoda, *in* Kaesler, R.L., ed., Treatise on invertebrate paleontology, Part R, Arthropoda 4, Volume 3: Boulder, Colorado, Geological Society of America (and University of Kansas Press), 277 p.

Kane, J.S., and Neuzil, S.G., 1993, Geochemical and analytical implications of extensive sulfur retention in ash from Indonesian peats, *in* Cobb, J.C., and Cecil, C.B., eds., Modern and ancient coal-forming environments: Geological Society of America Special Paper 286, p. 97–106.

Keller, G., 1992, Paleoecologic response of Tethyan benthic foraminifera to the Cretaceous-Tertiary transition, *in* Takayanagi, Y., and Saito, T., eds., Studies in benthic foraminifera: Tokyo, Tokai University Press, p. 77–91.

Sawyer, D.S., Buffler, R.T., and Pilger, R.H., 1991, The crust under the Gulf of Mexico basin, in Salvador, A., ed., The Gulf of Mexico Basin: Boulder, Colorado, Geological Society of America, Geology of North America, v. J, p. 53–72.

Taylor, J.C.M., 1990, Upper Permian-Zechstein, *in* Glennie, K.W., ed., Introduction to the petroleum geology of the North Sea (third edition): Oxford, United Kingdom, Blackwell, p. 153–190.

Guidebook
Barton, C.C., and Hsieh, P.A., 1989, Physical and hydrologic-flow properties of fractures, *in* International Geological Congress, 28th, Field Trip Guidebook T385: Washington, D.C., American Geophysical Union, 36 p.

Blackstone, D.L., Jr., 1990, Rocky Mountain foreland exemplified by the Owl Creek Mountains, Bridger Range and Casper Arch, central Wyoming, *in* Specht, R., ed., Wyoming sedimentation and tectonics: Casper, Wyoming Geological Association, 41st Annual Field Conference, Guidebook, p. 151–166.

Burchfiel, B.C., Chen Zhiliang, Hodges, K.V., Liu Yuping, Royden, L.H., Deng Changrong, and Xu Jiene, 1992, The South Tibetan detachment system, Himalayan orogen: Extension contemporaneous with and parallel to shortening in a collisional mountain belt: Geological Society of America Special Paper 269, 41 p. *[Note that Chinese names are commonly arranged family name first, followed by given name; in this example, the family names of the Chinese authors are Chen, Liu, Deng, and Xu. For references with Chinese authors, spell out the entire name.]*

Peirce, J.W., Weissel, J.K., and others, 1989, Initial reports, Ocean Drilling Program, Leg 121: College Station, Texas, Ocean Drilling Program, 1000 p. *[Include names of both co–chief scientists.]*
Shipboard Scientific Party, 1987, Site 612, *in* Poag,

C.W., Watts, A.B., et al., Initial reports of the Deep Sea Drilling Project, Volume 95: Washington, D.C., U.S. Government Printing Office, p. 31–153.

Twiss, R.J., and Moores, E.M., 1992, Structural geology: New York, W. H. Freeman and Company, 532 p.

Vogt, P., and Tucholke, B., editors, 1986, The western North Atlantic region: Boulder, Colorado, Geological Society of America, Geology of North America, v. M, 696 p., 11 pl.

Weaver, C.E., 1989, Clays, muds and shales: Amsterdam, Elsevier, Developments in Sedimentology, v. 44, 819 p.

6.1   Reference Style Guide for Paper/Assignments

Open-File Report
Alpha, T.R., 1993, Landslide effects: U.S. Geological Survey Open-File Report 93-0278-A (paper copy, 43 p.) or 93-0278-B (3 1/2" Apple diskette).

National Earthquake Information Center, 1992, Earthquake data report, August 1992: U.S. Geological Survey Open-File Report 92-0608-A, 458 p.

Proceedings from a Symposium or Conference
[Include year of conference if it differs from publication year.]
Baar, C., 1972, Creep measured in deep potash mines vs. theoretical predictions, *in* Proceedings, Canadian Rock Mechanics Symposium, 7th, Edmonton: Ottawa, Canada Department of Energy, Mines and Resources, p. 23–77.

MacLeod, N.S., Walker, G.W., and McKee, E.H., 1976, Geothermal significance of eastward increase in age of upper Cenozoic rhyolitic domes in southeastern Oregon, *in* Proceedings, Second United Nations Symposium on the Development and Use of Geothermal Resources, San Francisco, May 1975, Volume 1: Washington, D.C., U.S. Government Printing Office (Lawrence Berkeley Laboratory, University of California), p. 465–474.

Thesis
Wopat, M.A., 1990, Quaternary alkaline volcanism and tectonics in the Mexican Volcanic Belt near Tequila, Jalisco, southwestern Mexico [Ph.D. thesis]: Berkeley, University of California, 277 p.